ROSAMOND GIFFORD ZOO
at BURNET PARK

BARBARA SHEKLIN DAVIS

Foreword by Ted Fox, Executive Director,
Rosamond Gifford Zoo

Published by The History Press
Charleston, SC
www.historypress.com

Copyright © 2025 by Barbara Sheklin Davis
All rights reserved

Unless otherwise indicated, all photographs are courtesy of the Rosamond Gifford Zoo.

Front cover, bottom: Asian elephant Mali made history when she gave birth to twin boys, Yaad and Tukada, on October 24, 2022. *Photo by Camille Murphy. Back cover, top:* Giant Pacific octopus Ollie arrived at the Rosamond Gifford Zoo in November 2018. *Photo by Maria Simmons. Bottom right*: Twin Amur leopard cubs Milo and Mina were born on June 19, 2019, to parents Tria and Rafferty. *Photo by Maria Simmons.*

First published 2025

Manufactured in the United States

ISBN 9781467157889

Library of Congress Control Number: 2024945353

Notice: The information in this book is true and complete to the best of our knowledge. It is offered without guarantee on the part of the author or The History Press. The author and The History Press disclaim all liability in connection with the use of this book.

All rights reserved. No part of this book may be reproduced or transmitted in any form whatsoever without prior written permission from the publisher except in the case of brief quotations embodied in critical articles and reviews.

CONTENTS

FOREWORD

Welcome to the wonderful world of the Rosamond Gifford Zoo, a sanctuary of knowledge and nature nestled in the heart of Syracuse, New York. This history book invites you on a journey through time, exploring the evolution of an extraordinary institution that has become a cherished landmark in our community since its first iteration more than a century ago.

The Rosamond Gifford Zoo isn't just a collection of exhibits housing rare creatures from around the world—it's also a living testament to the enduring human fascination with the animal kingdom. The following pages recount the zoo's past from its humble beginnings to the vibrant and dynamic institution it is today—a steady progression fueled by vision, passion and unwavering commitment to conservation and education.

Established in 1914, the zoo has undergone a remarkable transformation, mirroring the evolving understanding of our relationship with the natural world. What started as a modest gathering of local animals has flourished into a world-class facility that houses a diverse array of elusive, exotic and endangered species. The Rosamond Gifford Zoo has not only provided a home for these remarkable creatures but also served as a beacon of education, fostering a deep appreciation for wildlife and the delicate ecosystems that sustain them.

The book, like the institution, is a gift and resource to a community that has embraced and engaged with the zoo throughout its existence—and stood by with steadfast support through the greatest of challenges.

It is a tribute to the dedicated individuals (from the early pioneers to our modern-day care specialists) who have shepherded the zoo and its animals while expanding the institution's impact. It is a celebration of the efforts of zookeepers, educators and conservationists who work tirelessly to ensure the well-being of the animals in their care and to inspire the next generation of wildlife enthusiasts.

The story of the Rosamond Gifford Zoo is a story of conservation, education and the enduring connection between humans and the awe-inspiring creatures with whom we share our planet. As you turn the pages, you will encounter stories of trials and triumphs, births and farewells and the countless moments that have left an indelible mark on the zoo's history. The Rosamond Gifford Zoo has weathered the passage of time and emerged stronger, more resilient and more committed than ever to its conservation mission and core principles.

So join us on a historical journey to see how far we've come and to envision how far we can go. We hope this book will provide greater understanding and appreciation of the work that has gotten us this far—and that the history recounted here will inspire a new generation of conservationists to continue the vital work of protecting the natural world for centuries to come.

Ted Fox
Executive Director
Rosamond Gifford Zoo

INTRODUCTION

Over seven hundred million people visit the world's ten thousand zoos each year. It has been calculated that in the United States, more people visit accredited zoos every year than attend professional baseball, basketball, football and hockey games. Worldwide, the popularity of zoos cuts across all ideological boundaries. Zoos are incredibly popular in capitalist, communist and third world countries alike.

But zoos are also controversial. Adrienne Stroup, a student who researched local zoos, wrote in 2012,

> *On some level, zoos are no different from conventional museums in how they serve the public, operate within their communities, and network between other zoos. Like museums, zoos provide an educational and entertaining experience for the general public. In fact, zoos* are *museums, and the only difference between them and other museums is that their collections are living.*

While there are those who see zoos as modern Noah's arks, saving species that would otherwise go extinct through habitat destruction, others, like the World Animal Foundation, argue that the containment of animals is inherently cruel and that "zoos are bad because animals are forced to live in unnatural, stressful, boring environments, leading to a lack of mental and physical stimulation. They are removed from their natural habitats and confined to small, limited spaces and often forced to perform tricks or entertain visitors."

People in Syracuse had long wanted a zoo. As early as 1898, an anonymous donor offered "to build and stock a zoological department in the proposed park to the South end of the city if the city will take steps to fix up the lands for park purposes." The local newspaper reported that "the plan meets with enthusiastic approval, and all that the city has to do in the matter is to set aside the lands for park purposes. The only expenses will be for improving them." The donor was certain "that a good sum for this purpose can be raised by public subscription," but the plans did not materialize.

Not to be deterred, others stepped forward to create a small collection of animals designed to intrigue and entrance the public. Some donated money to the cause. Others were willing to donate animals. In 1946, a Mattydale resident personally delivered a five-hundred-pound bear. The Manlius Town Board donated six cygnets from its swan pond. Three peacocks were donated by a stock farm in Rhode Island. The zoo acquired other animals in less acceptable ways. In 1957, the zoo announced the impending arrival of one hundred male monkeys "fresh from the African jungles"—but only "if a searching party has luck during the winter capturing that many of the animals to fill city's order." In 1966, the county's parks commissioner announced that trappers were "sloshing around the treacherous Amazon River jungles" in quest of three rare red-faced uakari monkeys. "They haven't been captured yet," he said, "but we're going to get them."

Despite the good intentions of the early creators of the Burnet Park Zoo, it eventually became clear that goodwill was not enough to sustain a quality zoo that took good care of its wildlife. Eleven small animals were killed in a nighttime raid by vandals in 1970. In 1974, two teens broke into the zoo and killed and injured about forty animals. But the worst case of vandalism in the zoo's history occurred in 1975, when two youths, aged twelve and nineteen, enraged the community by kicking and stabbing twenty-seven animals in the children's zoo to death. "The city's financial position and the break-in fueled public debate over the future of the zoo."[1] Critics said that the main building housed too many animals of incompatible species in conditions that amounted to animal cruelty. A local biology teacher described the zoo as "a building of 10,000 animals, most of them flies." Outraged by the vandalism and disease at the zoo, critics called for a huge increase in funding or closure. But the City of Syracuse did not have the funding required to do the needed repairs and upgrades. Onondaga County had to step in to save the zoo.

For most of the zoo's early history, there had been tension between those who saw it as people-centered entertainment and those who saw it as animal-centered management and conservation. The zoo went through several

The Burnet Park Zoo was always a favorite destination for school field trips, as evidenced by this 1930 photo of a class visit. *Courtesy of the Onondaga Historical Association.*

iterations before emerging as the facility it is today, attractive to hundreds of thousands of visitors annually and home to seven hundred animal species ranging from the smallest (milkweed bug) to the largest (Asian elephant). Like most zoos, the Syracuse zoo began as an exotic collection or menagerie of animals designed for human self-aggrandizement and viewing pleasure. It subsequently evolved to take on the challenge of educating humanity about the importance of the animal world and conserving the world's most endangered species. As humans increasingly encroached on and destroyed animal habitats, zoos embraced a mission to care for, protect and preserve the animals on the planet and hoped, through research and education, to inspire the development of a societal ethos of care and compassion for all creatures. William G. Conway, director of the Wildlife Conservation Society and president of the American Zoological Association, has said, "The justification for removing an animal from the wild for exhibition must be judged by the value of that exhibition in terms of human education and appreciation, and the suitability and effectiveness of the exhibition in terms of each wild creature's contentment and continued welfare."

Beginning with a little elephant on a cake at a dinner party at the beginning of the twentieth century, the zoological garden that would become the Rosamond Gifford Zoo of the twenty-first century has been built on pachyderms. It was the plight of Oodles the elephant that first called public attention to the need for better elephant quarters, and it was little Siri doubling her size that forced the zoo to undergo its first expansion. Elephants enthralled the community through the highs of successful births and the lows of tragic deaths. While the zoo encompassed a tremendous range of creatures for which it cared, it was its work with the world's largest land mammals that captivated the most attention and exemplified its excellence.

As it embarks on its second century, the Rosamond Gifford Zoo has sharpened its focus and redefined its mission. From things as small as banning plastic straws in its food service areas to working with businesses to enhance their understanding of the zoo's impact, the zoo has embraced the challenge of the twenty-first century: how to engage, inspire and educate community members on conservation and sustainability. The zoo seeks to teach people about their potential to have a positive impact in their community and be a force of good for planet Earth. To advance this conservation education mission, the zoo strives to provide a quality venue and services that are accessible to all—by improving the grounds, diversifying the workforce, expanding community offerings and engagement and making its practices more sustainable, in order to create meaningful and lasting connections between people and the natural world.

Chapter 1

IN THE BEGINNING

What do typewriters have to do with a zoo? Well, the fact is that without them, Syracuse's Rosamond Gifford Zoo would not exist. The Smith brothers, Lyman, Hurlburt, Monroe and Wilbert, were gun manufacturers. One of their employees, Alexander Timothy Brown, was an inventor. In 1887, Brown created a typewriter that was the first to include both upper- and lowercase letters. The Smiths decided to manufacture it. The typewriter would be the Smith-Premier No. 1, patented in 1889 under patent number 411,421.[2]

> *The Smith Premier was the most advertised and successful double keyboard typewriter of its time....Model 1 is distinguished from later models by its exquisite, nickel-plated, embossed pattern on the frame....*[Its] *unique design of transferring motion from the keys to type bars gives a very smooth and responsive touch for the typist. It sold for $100, in line with many keyboard typewriters of the day. In comparison, a horse-drawn carriage cost around $60.*[3]

"These 1st models sold well. So much so that the Smiths exited the firearms business to concentrate exclusively on writing machines. They erected a new factory to support the booming business in Syracuse, NY on the corner of Clinton and Onondaga Streets" in downtown Syracuse and, in 1926, merged with the Corona Typewriter Company.[4] "The combination of LC Smith's durable office typewriters and Corona's popular portable machines made the new firm an industry leader and helped them to remain profitable even during the Great Depression."[5]

Left: The Smith Corona typewriter was a link between a wealthy benefactor and one of the first efforts to create a zoological garden in Syracuse. *Courtesy of the Onondaga Historical Association.*

Right: Lyman Smith, who made his fortune in typewriters, was one of the earliest philanthropists to offer to create a zoo for the city. *Courtesy of the Onondaga Historical Association.*

Lyman Smith was a philanthropist. He donated funds to build the Lyman Cornelius Smith College of Applied Science on the Syracuse University campus and then gave additional monies to erect Smith Hall, which later housed the College of Visual and Performing Arts. He was a major financial booster of Syracuse rowing, and on the opposite side of the country, he funded the Smith Tower in Seattle, Washington. The Smith Tower, one of the world's first skyscrapers, was hailed as the tallest office building in the world outside of New York City.

As recounted by local historian Kihm Winship on her *Skaneateles* blog, the Smith family lived at 804 James Street in a nineteen-room turreted mansion with stained-glass windows. Lyman was a horticulturist and had greenhouses filled with orchids. He had a conservatory that included an aquarium, an ornate Turkish smoking room and a collection of rare plants. The Smith home was named Uarda, which was spelled out in flowers in the front yard. The word *uarda* is of Arabic origin and means "a bouquet of flowers."

> *In the summer of 1904, L.C. Smith suggested to his head gardener, Joseph Kenney, that a large floral elk would be just the thing for the front lawn of Uarda. The finished elk, of wood lath, wire, potting soil, moss and flowers, weighed 1,400 pounds and required 12 men to carry it from the hot house to the front lawn; the antlers had a five-foot spread.*[6]

Top: The Smith mansion had a greenhouses for orchids, an aquarium, an ornate Turkish smoking room and a collection of rare plants. *Courtesy of the Onondaga Historical Association.*

Bottom: Flora Smith, Lyman's daughter, was raised to be philanthropic and shared her father's love of all things extravagantly floral.

Smith's daughter, Flora Bernice, grew up in luxury but was also taught to be generous. The household staff numbered twenty-two. At the age of five, Flora christened a ship named for her father, the *Lyman C. Smith*, built for the United States Transportation company for use on the Great Lakes. Family and friends traveled to Michigan in a special train car for the occasion. Flora learned to be philanthropic. Each year at Christmas, she gave a doll or toy to every patient in the children's ward of the Syracuse Women's and Children's Hospital. Aptly named, she shared her father's love of all things extravagantly floral. To celebrate Charles Lindbergh's 1927 flight across the Atlantic, she had the Uarda gardeners create a floral replica of the *Spirit of St. Louis* on her front yard. The local newspaper noted, "For several weeks now it has been admired by motorists and pedestrians on James Street."

The Smiths entertained lavishly. One of their most notable gatherings occurred in 1902, when the crown prince of Siam visited Syracuse. Prince Vajiravuda wanted to come to Syracuse because his Siamese tutor had taught him how to type on a Smith Premier typewriter, the first to use a Siamese alphabet keyboard. Lyman Smith hosted a large dinner party for the prince at his home.

The mansion was trimmed with roses and mums; the guests, all men, listened to Oscar Kapp's orchestra and sat down to a 30-plate luncheon. The menu included Lynhaven Bay oysters, Steinwein Bocksfeutel (a German white wine), clear green turtle soup, Amontillado (a Spanish sherry), a selection of celery, olives, cucumbers, radishes and almonds, filet of sole

Madeleine, sweetbreads pique, French peas, stewed Maryland terrapin, a punch "surprise" served in a miniature brass coal pail, canvas back duck, Chambertin (a Pinot Noir of Burgundy), salad fantasie, biscuit tortoni, petits fours, liqueurs, bonbons, fruit, cigars and coffee. There was also ice cream in a little box surmounted by a white elephant, the symbol of the Siamese royal family.[7]

Little did anyone realize that the little elephant was a portent of the future. On March 1, 1900, a notice appeared on page 7 of the *New York Tribune*:

Lyman C Smith, the wealthy typewriter manufacturer, has offered to give the city a magnificent zoological collection, which will include rare animals from the tropics. An appropriation of $10,000 has been placed in the city budget for the improvement of Onondaga park, where the gardens will be located. Mayor McGuire recently had a talk with Mr. Smith, in which the latter said that if the park is improved, he will do something handsome in

Left: The Prince of Siam came to Syracuse because of a typewriter and was hosted royally by the Smith family. *Courtesy of Wikimedia Commons.*

Right: William W. Wiard, president of the Syracuse Chilled Plow Company, was the man whose enthusiasm for the project made the very first zoo possible. *Courtesy of the Onondaga Historical Association.*

> *the way of a zoological collection. Mr. Smith is said to have in mind a collection which will include swans, peacocks and other birds. An artificial lake will be made. Mr. Bishop, Superintendent of Parks, has been instructed to go ahead with the work of building roads and getting the park in shape for the "zoo," which will be done this year, and the summer following will probably see the park open to visitors.*

Sadly, Lyman Smith's plan for the zoo never advanced. But then another wealthy local businessman stepped up. In 1915, William W. Wiard, president of the Syracuse Chilled Plow Company, traveled on business to Toronto, bringing his wife and son, William Jr., along. While Wiard made business calls, his family visited the Toronto Zoo, which his son described as "the greatest show on earth." William Jr. insisted that his father come to see the zoo, a visit that a local newspaper described as "destined to make Syracuse history." Wiard Sr. said that the Toronto Zoo made an impression that "got under our skin," and when he returned home, he set to work making plans to bring a zoo to Syracuse.

At a dinner at his home attended by Syracuse mayor Edward Schoeneck, Wiard made his case for a zoo for the city. The response was enthusiastic, with Mayor Schoeneck declaring to Wiard, "You are the man to head the Municipal Zoo Commission." The commission, with Wiard as chair, set two conditions for the zoo: the animals and birds had to be provided at no cost, and the municipality would provide housing and maintenance for them.

It was the support of Syracuse mayor Edward Schoeneck that led to the creation of the first municipal zoo in the city in 1914. *Courtesy of the Onondaga Historical Association.*

The Zoo Commission was really a one-man show, but Wiard's enthusiasm for the project was limitless. He purchased a black bear named Geronimo from the Catskills and added a pair of foxes. The animals were housed in a large red barn on a four-acre parcel of Burnet Park. The barn doors were swung open to the public in the morning and closed at night.

Syracuse's next mayor, Louis Will, son of the founder of the Will & Baumer Candle Company, was even more sympathetic to the zoo project and found funds to build a fenced enclosure. Wiard then supplied white-tailed deer. Other animal lovers stepped forward, and in short order, three buffalo, six Adirondack deer and a pair of elk had been donated. H.H. Franklin, the car magnate, provided many birds.

In the 1940s, Vera E. Klock, a Syracuse schoolteacher, wrote a history of the zoo. In describing its origins, she noted,

> [It] *didn't have a law to create it; it didn't have even a salary to reward it. Like Topsy, it just "grewed" from a group of philanthropic gentlemen who became interested in the idea of a zoo for Syracuse. They made it their business to secure from others donations of money or animals with which to start.... The project caught on quickly. In 1913 and 1914, donations of any sum or any animal were eagerly accepted. The collection was housed in an old barn.... Because of the insufficient quarters, it was not possible to keep any animal which needed special care in the winter. However, the first zoo could boast of raccoons, bears, deer, foxes, peacocks, parrots, pheasants and owls.*

Klock gave a complete accounting of the donations. In October 1914, the collection included two black bears, two raccoons and two gray foxes from William Wiard; a white buck and doe from Louis Wiard; two Canadian black bears purchased by E.I. Rice at Star Lake in the Adirondacks; a seven-year-old Florida alligator from Mrs. W.W. Winchester; four white swans from Louis Wiard and four from President Wiard, who also gave seven Canada geese; thirty white wild ducks, a dozen gray squirrels, a dozen prairie dogs and six wild ducks; a pair of Nicaraguan parakeets from Lieutenant Everson; and a dozen pheasants (ten ring-necked, two golden) and three large owls from Forman Wilkinson. The waterfowl were kept in the pond in Onondaga Park as the zoo had no pond, and a sand pit was built for the prairie dogs. But it was clear that more accommodations were necessary, so the zoo committee planned to buy new cages for the rapidly increasing population. The committee also voted to purchase a buffalo. "People would like these huge, rare

William W. Wiard's successor, Louis Will, son of the founder of the Will & Baumer Candle Company, found funds for fenced enclosures at the zoo. *Courtesy of the Onondaga Historical Association.*

creatures, they were sure, and it could be kept in an outside enclosure," wrote Klock.

But care at the zoo was substandard. Geronimo died from being underfed, and the foxes went blind from being kept in the dark for too long. Providing proper care for the animals was a challenge from the very beginning. In 1914, three raccoons were ordered and shipped but did not arrive. After some time, they were found at the trolley station in bags inside boxes. Two had been smothered, but the third was brought to the zoo.

Word was then received that Alex T. Brown was giving three buffalo to the city. One, named Jerry Rescue, was to come from the Midwest, and the other two were to arrive from Massachusetts as soon as they were released from quarantine. It was also suggested that the zoo start a squirrel collection with twelve each of gray, red, black, white, orange, fox and flying squirrels. To provide for all these animals, as well as a hoped-for deer herd, an appeal for funds was made to the Syracuse Common Council.

The year 1915 was a historic one for the zoo in terms of animal acquisition and funding. A turning point occurred when the Whitcomb Zoo, a traveling menagerie, came to town. It had about one hundred different varieties of animals and birds and attracted many visitors while on display in a store opposite the Empire Theater on South Salina Street. The menagerie's owner, George Whitcomb, approached William Wiard and offered to sell his stock to the city for $10,000. The proposal was initially rejected because the price tag was too high, but eventually, the owner and Wiard found an acceptable solution. The animals and birds would be exhibited under canvas at Burnet Park for one month to determine whether attendance was sufficient to justify the establishment of a zoo. If so, Wiard agreed to pay $3,000 for the stock.

Thousands of people came to see the animals over the next several weeks. The estimated twenty thousand visitors so taxed the trolley system that new switching needed to be installed. William Wiard raised the necessary capital from his personal funds and from contributions by many of the prominent citizens of Syracuse. Klock reported that "the dream of W. Wiard and his colleagues was now a firm reality, a growing part of our city's highlights."

April 15, 1915, was a memorable day for the zoo in terms of funding. The Syracuse Common Council authorized and directed the City of Syracuse

> *to issue bonds in the sum of $10,000 for the following specific municipal purpose, to wit: to provide for the construction of buildings or structures*

> *suitable for the housing of animals in the zoological exhibit belonging to the city of Syracuse and boarded in Burnet Park in said city and for the grading and improving of the land around the aforesaid buildings, such construction to be under the control, supervision and direction of the Park Commission in end for said city created pursuant to provisions of Chapter 5s6 of the law of 1906 of the State of New York.*

The minutes record that "there was no objection to the ordnance of a $10,000 sum so it was accepted."

Klock wrote, "Now could be the dreams fulfilled, permanent buildings, adequate food, landscaping, all that which would make the zoo of the City of Syracuse equal to that of any other city." Other animals were procured through donations by prominent local citizens and were kept in small, barred cages for public viewing purposes. Mayor Louis Will gave the zoo two white-tailed deer, four white swans and two elk; Charles Estabrook donated twenty monkeys; and the *Herald Journal* gave the zoo a pony named Newsy.

Wiard made public the plans for enlarging the zoo to occupy several acres in the huge Burnet Park. As Klock described it,

> *The Commission planned to use the $10,000 to lay out a miniature city of buildings, walks and drives. The entrance was to be at Milton Avenue a few feet from the then Solvay car line. A broad walk several city blocks in length would cut directly across the park from Milton Avenue to Grand Avenue, along the line of the iron picket fence separating the park from the Syracuse State Institution for the Feeble Minded. To the left of the walk, as you come from Milton Avenue, a large bear pen was already under construction on the side of a hill. A little further along, a large bird cage was to be erected in the midst of a clump of trees. Among the trees along the walk are to be the cage houses for the smaller animals. At the end of the walk, near Grand Avenue, is a deep ravine of at least three acres. Here was to be the runways for deer, elk and buffalo.*

So the zoo began to be built. First came the stone dens, then the aviary, then the renovation of the barn. At about this time, Syracuse had its first brush with an elephant. Flush with the success of the menagerie acquisition, the *Syracuse Herald* declared that the zoo needed an elephant because no zoo was complete without one. The paper began a campaign to raise the $1,500 cost of a pachyderm. When William Wiard was approached for his

A bird cage and a bear pit were prominent features of the zoo in its earliest days.

opinion of the venture, he at once donated ten dollars in his son's name. It was then decided that no one could give a donation except in a child's name. It was to be the children's elephant. With the sanction of the zoo commission, the *Herald* started an elephant campaign, encouraging every child to give, no matter how little. It published a daily listing of those who had contributed amounts from a penny to ten dollars, declaring, "These children will make our cities' best citizens." It was then decided to give elephant naming rights to the school that donated the most money. The Merrick School won the competition, and the elephant was to have been named Merrick, "Merry" for short. Unfortunately, World War I interfered with the fundraising, and the $1,500 goal was never achieved.

The elephant acquisition was not the only casualty of the war. Klock reported that interest in the zoo declined "as we were swept into the war." A local paper reported that food rations for the animals "have been curtailed and some food is not even possible to get." The city's purchasing department requested permission to acquire food without going through the usual competitive bidding process. "They found it better to search around to get what was needed. Mr. Healey, head keeper, says now they are considering growing their own vegetables for the animals since it takes so much and is so hard to get," Klock reported. She noted, "There were no new buildings, very few donations of money or animals for the zoo, for the zoo was at a standstill as was everything else until the war was to be won." There was no steel to make repairs to animal cages and even with the indomitable Mr. Kallfelz at the helm of the Zoological Commission, the city comptroller's office concluded that "time will tell whether our zoo will go on to newer greater heights or remain static after the war."

The birth of a bear cub at the zoo in 1923 was a cause for jubilation. The cub was believed to have been born in February, when squeals were heard

Spider Monkey and Buster, New Arrivals at Zoo Are Happy as Spring Appears

Jocko, the spider monkey, entertains Miss Cynthia Dayan, 602 East Genesee Street, in Burnet Park Zoo with Buster, the bear cub, assisting at the left.

A newspaper drawing shows a zoo visitor being entertained by Jocko the spider monkey and Buster the bear cub. *Courtesy of the Onondaga Historical Association.*

in the bear cave, but the parents were hibernating, so the birth could not be verified. Despite attempts by zoo superintendent Dan Hanely to entice the parents out with fish morsels, it was not until March that the cub appeared. The *Syracuse Journal* reported that "congratulations are in order for Mr. and Mrs. Bruin, address, Burnet Park Zoo, who are now the proud parents of a tiny bear cub, the first born in Syracuse since the time the Indians swarmed the hillsides." The cub was said to resemble a "baby chipmunk more than he does a really truly live bear, but he gives promise of growing up." Schoolchildren were invited to submit ideas for Baby Bruin's name, and Buster was chosen.

Alas, Baby Buster's life was brief. In July 1923, his mother became ill and died from a stomach disorder, which she passed on to Buster, who died a week later. "The shadow of death and unhappiness has thrown its pall over the animal kingdom in Burnet Park," wrote the *Journal*.

> *Children are struck with the quiet and sadness which hangs like a dark cloud over a place usually so full of life and noise. The monkeys and the baboons have ceased their chattering and sit listlessly in their cages, scratching their little heads and pondering on the catastrophe which has befallen their neighbors. Each little animal desperately asks its mate, "Isn't there something we can do for the bears?" but as it always happens, there is nothing.*

The reaction to Buster's death demonstrated how the zoo had become an important part of the Syracuse community, as illustrated in a 1931 newspaper cartoon.

Soon there were enough animals to justify a new building. In 1929, the Syracuse Common Council allocated $50,000 for that purpose. A March 9, 1930 article in the *Syracuse Herald Journal* described it as

> *an attractive, ornate structure of brick, 150 feet long and 50 feet in width. At one end will be the monkey cage, at the other provision will be made for the housing of the birds in the Zoo. Necessary warmth for both in cold weather will be assured by partitions, adding to the efficiency of the heating system. There will be approximately 30 outside cages, that is, cages open to inspection from both without and within. This new building will be the major unit of the Zoo, for the present at least. It supplements, or if you prefer, completes the layout which embraces sturdily built enclosures for buffalo, deer and elk, dens for wolf, fox and prairie dog, and the flying cage and pond for ducks and birds.*

Peter Kallfelz, owner of Kallfelz Bakery and an animal lover, often complained the city wasn't buying enough food for the zoo's animals. He became a financial backer of the zoo and purchased animals, donated food, took animals home to raise them and helped zookeeper Dan Hanley care for the creatures. Klock noted that "from the 1930s to the present day, we find the name of Peter B. Kallfelz on almost every donation which the zoo has received. Through financial reverses, Mr. Wiard was no longer able to be the foster father to the city zoo. Mr. Kallfelz took up where Mr. Wiard left off, buying food and animals."

In May 1932, the zoo announced that Kallfelz had placed an order for two lion cubs with a wild animal importer in Texas. "The cubs are 7 months old and we expect that they will more than make up for the lion we lost several months ago," said Commissioner of Parks William A. Barry.

A newspaper cartoonist sketched Burnet Park Zoo animals and personnel for a picture titled *Going Places and Seeing Things. Courtesy of the Onondaga Historical Association.*

The new brick zoo building constructed in 1929 had thirty outdoor cages, open to viewing from both inside and outside. *Courtesy of the Onondaga Historical Association.*

Kallfelz was unstoppable. An article in the *Herald Journal* in October 1932 reported,

> *For the first time in the history of the Burnet Park Zoo, the main building of the menagerie was filled to capacity today with the arrival of 12 new animals and birds, all gifts of Peter J. Kallfelz. Among the shipment were two baby hyenas. Two Sarus cranes from India, the largest birds of their type and said to be rare in this country, were included in the new arrivals. Two pigtailed monkeys from India were placed in a cage next to Leo the chimpanzee. A capybara of South America, the largest of all rodents, received a cage by himself and crouched there motionless, not even blinking an eye at the spectators. The rest of the animals watched with curiosity the unloading of the new arrivals, which included two gray breasted whistling tree ducks from South America, two arid toucans, also from South America, and a coatimundi from Central America.*

That year it was reported that 250,000 people had visited the zoo over the summer. Kallfelz was elected chairman of the Municipal Zoo Commission in 1933, succeeding William Wiard. He continued to work hard on behalf of the zoo. In 1933, William A. Barry, Syracuse commissioner of parks, said that the Burnet Park Zoo "is rapidly becoming one of the best in the country for a city the size of Syracuse." He said that "the small animal section was very well represented with ten rhesus monkeys, two mangabey monkeys, two green monkeys, two spider monkeys, two capuchin monkeys, one chimpanzee and one Arabian baboon."

Barry described the bird collection as containing "pheasants of every description, collected from China, Europe and South America, as well as the United States. Also in the collection are a king vulture, aerial toucans, red-

The zoo was a very popular place to visit. In 1932, a quarter of a million Syracusans came to see the animal collection. *Courtesy of the Onondaga Historical Association.*

The zoo's bird collection was large and diverse and included such species as a toucans, turtledoves, macaws, cranes, parrots, peacocks and swans.

billed toucans, turtle doves, tree ducks, red macaws, tarus cranes, parrots, an American eagle, peacocks, swans and Canadian geese." Rounding out the collection were the larger animals: one bison, two elk, two red deer, two laughing hyenas, three lions, one tiger, one jaguar, four leopards and five bears. "Altogether," he said, "there are 237 animals housed in the zoo." In 1934, Kallfelz gave the zoo a pair of rare Chinese white peacocks, a male and a female.

A bear became the newest inhabitant of the zoo when he was brought to town to perform in a downtown store window. "Needless to say," writes Krock, "Geronimo got no further but was promptly added to the zoo, joining Josh and Ned, the two bears already living in Burnet Park."

There was no doubt about the zoo's popularity with the public. A 1934 article in the *Syracuse Herald* reported that more than ten thousand people came to the zoo in one day to see Petey, the nine-day-old son of the zoo's monkeys Jocko and Minnie. But the zoo's collection was haphazard at best. Fibber McGee was a five-hundred-pound bear who was donated and personally delivered to the zoo in 1956 by his owner, who lived in Mattydale. A chimpanzee was introduced with the headline "Mary the Moocher of New York Makes Her 'Social Debut' at Burnet Park Zoo." The reporter's lack of sensitivity to animal behavior was clear in the accompanying article,

Housing animals in cramped quarters—despite its ill effects, often evident to visitors—were the status quo for zoos of the time. *Courtesy of the Onondaga Historical Association.*

which reported that when Mary was introduced to Leo—the chimpanzee "who one day will acknowledge Miss Mary as his bride"—Leo showed his teeth and uttered "a veritable flood of chimpanzee chatter" and "insisted on Miss Mary entering his cage at once. When this was not to occur, he attempted to tear down the cage."

Although philanthropic and well-intentioned, these early animal supporters were neither knowledgeable nor successful in keeping their charges healthy or long-lived. The lion's death was succeeded by the deaths of a buffalo and Geronimo the bear. Vandals claimed the lives of other animals: in 1946, a white swan was shot and two deer were killed. In 1956, two swans were found hanging from a tree. In 1965, a Sebastopol goose was killed, its body tossed into a pond and its eggs thrown at other animals. That same year, a five-month-old bear cub was poisoned with mothballs. Then the zoo's big cats began to die. First to go was Sabre, a male African lion, followed by the lioness Burnetta; Lillie, the Bengal tiger; the black panther Puma; a clouded leopard; and an ocelot. By the end of 1968, more than half of the zoo's cat collection was dead. In 1970, vandals slaughtered eleven small animals, and, in 1975, twenty-seven animals were stomped, stabbed or crushed to death.

Chapter 2

YEARS OF DECLINE

People on unemployment and forced leisure went to the zoo regularly during the Depression; there was no admission charge. But the zoo was on a downward trajectory. "Syracuse's tax base [had] started to shrink and financial support for the zoo began to erode."[8] The facilities were substandard. One visitor recalled "buildings with cages on both sides and not much room for an animal like a lion to turn around in. It always smelled, a real bad odor." The zoo became better known for its smells than its sights.

The zoo had no plan for animal acquisition and little concern over appropriate accommodations. In 1955, Alexander F. Jones, executive editor of the *Syracuse Herald Journal*, wrote a scathing editorial saying that "anyone walking in the Burnet Park Zoo must have a cast-iron stomach. It is not that clean animal smell. It is a nauseating stench." He did not lay the blame on park officials or keepers but rather on the city council that "only sees fit to appropriate from $25,000 to $30,000 annually to maintain the Burnet Park Zoo." With this limited budget, he wrote, "there are no facilities for ventilation and cleaning equipment is not modern" and "the animals themselves are obviously not in the best of condition." The only purpose of a zoo, he stated, "is to educate visitors, particularly children, on how wild animals look and act. The animals are supposed to be well fed and well housed so visitors can see them under the best conditions." The best that can be said for animals in the Burnet Park Zoo, he concluded, "is that they are kept alive."

Left: This souvenir zoo button depicting "see no evil, hear no evil, speak no evil" inadvertently exemplified attitudes at the zoo in its earliest days.

Right: The whale in the Children's Zoo was initially a source of delight but within a few years came to symbolize the zoo's decay.

As a result of the editorial, efforts were made to improve the animals' living conditions and update the grounds. A Children's Wonderland was built, with miniature structures housing a variety of farm animals with which children could interact. Funding for the project came from the Rosamond Gifford Charitable Corporation. "During her lifetime, Miss Rosamond Gifford had a tremendous interest in farm animals of all kinds," said Francis A. Feil, trustee and treasurer of the foundation, in an April 14, 1957 article in the *Syracuse Post Standard*.

> *She devoted a great deal of time and effort to the care of these animals. The trustees of the corporation, aware of her devotion, and with enthusiasm for Mayor Mead's ambitious plans for the modernization and expansion of the Burnet Park Zoo, approved a $40,000 gift for the construction of Children's Wonderland. This major step for improving the Zoo will provide educational and recreational benefits for the children of this community as well as adults.*

The local newspaper enthused about Wonderland:

> *Have you ever walked into a whale or watched Mary's lambs scamper and romp at their mother's side or had a close, close look at a skunk? These are some of the thrills awaiting your youngsters at the Syracuse Zoo. The*

> *Wonderland has a fresh sparkling look due to a facelifting this year. Clean new blacktopped walks wind through the attractively landscaped children's attraction. Bridges cross small streams of water and an aluminum shelter of modern design provides shade and there are some modernistic shapes for the kids to sit, climb or rest on.*

Additional funds were set aside by the city to renovate the bear den and construct Monkey Island. Still, visitors complained about the odor and general lack of cleanliness in the exhibit areas. A 1950s editorial headline said that the Burnet Park Zoo "rhymes with phew."

Within a couple of years, the Children's Wonderland needed repair. The whale's tongue was patched with tape, its upper lip was torn and the aquarium inside it was empty. Funding was difficult to obtain. Vandalism continued to be a problem. Zoo property was destroyed, and animals were released from their cages, stolen or killed. The housing for the animals was far below standard. The monkeys, for example, were housed in cages ten times smaller than the minimum size required by state and federal animal welfare regulations, and their cages were placed opposite the cages of the lions, the monkeys' natural predators. The lions' cages were also far too small.

In 1963, local insurance agent Greg Gualtieri accused zoo officials of running a "circus." Gualtieri proposed a $5 million city-owned natural habitat zoo on five hundred acres, on which animals could roam free, with no bars or cages. Gualtieri had traveled around the world visiting zoos and taking thousands of photographs to illustrate how this would work. He formed an organization, the Zoological Commission, to investigate the possibility of purchasing property north of the city and establishing a new state-of-the-art zoo and proposed a bond issue to cover the costs of the project.

Gualtieri worked with Sargent, Webster, Crenshaw and Folley, a Syracuse architectural and engineering firm, to develop the landscape, engineering and architectural design for his proposal. A table model of the park, prepared by the architects, was on display. The complex proposal included a "prey/predator"-type zoo, a "safari" restaurant, an aquarium, botanical gardens and, eventually, a natural history museum. "What we need is seed money," Gualtieri said. He noted that the design for the complex could be adapted to local desires and funding. "You have to look at it as an investment instead of an expense," he commented, adding that the plan would need a minimum of four hundred to five hundred acres but could cover up to one thousand.

According to a newspaper account, Syracuse's Mayor Walsh and Onondaga County executive Mulroy were on board with the project. Ultimately, though,

Businessman Greg Gualtieri's *(far right)* bold and brilliant vision for a zoo without cages was neither appreciated nor realized in his day. *Courtesy of the Onondaga Historical Association.*

City of Syracuse and Onondaga County officials were divided on the issue, agreeing only that the zoo should remain in Burnet Park.

The zoo purchased Oodles, an African elephant, in 1967, despite lacking appropriate housing for her. The inadequacy of Oodles's quarters was apparent even to a little girl who wrote to the newspaper saying, "We think that there should be a new zoo, because Burnet Park Zoo is all dirty and they have a little cage, and Oodles the elephant needs her feet taken care of, and she needs to be hosed too. Get a good zoo or plastic animals. I am nine years old and I love animals." The Humane Society reported that Oodles was housed in a facility "hurriedly constructed by destroying cage space already occupied by other exotic animals in the already overcrowded main building." It went on to say that Oodles's enclosure "proved too small for a growing elephant," and even after further expansion, "it still falls far short of the area required by a still small, still growing pachyderm." The Humane Society charged the zoo with running "an animal jail."

The inadequacies of Oodles's quarters were of little import to the two society ladies pictured in a 1967 newspaper alongside the small elephant's trunk, showing between the bars of a very small enclosure. The caption read: "It's the image of the GOP." A report by the Humane Society painted a very different picture, decrying "a facility that was hurriedly constructed by destroying cage space already occupied by other exotic animals in the already overcrowded main building." The report further cited Oodles's "grave foot complications" from poor drainage of urine in her stall and "a visible, repulsive dermatitis" on her forehead and trunk due to inadequate water for washing.

A 1967 article in the *Herald Journal* reported that Gualtieri's Zoological Commission was supportive of a city/county recreation study, which would include study of a new community zoo. The commission, which had fifty members, also said it was pleased with the purchase of an elephant for the existing zoo, although Gualtieri did note it was fortunate that the elephant was a baby, "giving a leeway of several years in finding adequate quarters for it." The elephant was housed in a small room in the main zoo building. But Gualtieri was urging the creation of a zoo of the habitat type, where animals could be seen in their natural surroundings. He suggested that a new zoo elephant house could also contain a hippopotamus, a rhinoceros and a giraffe. That same year, County Executive John Mulroy proposed a zoo on an eighty-acre site on the southwest shore of Onondaga Lake.

By 1972, it was clear that Oodles could no longer be kept in Syracuse, and she was sent to Lion Country Safari in Georgia, where she would have "acres of open grassy space and the presence of other elephants." The lack of an elephant weighed on the minds of the zoo's leadership, and not even a year after Oodles was relocated, another elephant was purchased. Siri was not yet five years old when she arrived, but just as the space was inadequate for Oodles, it was not fit for Siri either. Ironically, Siri, whose name means "to be free," was the zoo's original Asian elephant. She came to Syracuse from Chicago's Lincoln Park Zoo. For many years, Siri was the only elephant at what was then the Burnet Park Zoo, and she developed social skills with people instead of with other elephants.

On the Fourth of July 1975, Director Gray announced plans for an "outstanding zoo" with a price tag of $3.2 million. The proposal by the Friends of Burnet Park Zoo involved a "complete revamping" of the facility. The Friends, described as a "help organization," had decided "to pick up the ball and run with it," the director said. The Friends' plan was, in part, a reaction to the slaughter of two dozen small animals and the maiming of others in the zoo's children's section, Gray said. The Friends' plan was one of a number of proposals aimed at either enlarging and improving the

Burnet Park facility or building a county zoo, possibly in Lysander at a cost of $17 million.

The renovation cost of $3.2 million was "dirt cheap" when compared with the Lysander plan, Gray said, noting that security would be improved and the six-foot perimeter fence would have a patrol track around it, in contrast to the extant exterior fence, which could not be patrolled from the outside. He explained that each animal "would be in its own night stall" in the planned development, as opposed to the existing situation, in which "every animal outside the main building is out in the open." The grounds on the sloping site overlooking downtown Syracuse "would be landscaped 1,000 per cent better" to "make the grounds alone second to none." There would be "a complete animal kingdom" at the zoo, with space and facilities for insects and an aquarium. The effect hoped for "is to turn it into a complete educational and aesthetic experience."

"Money won't be easy. That's where the Friends of Burnet Park Zoo come in. The group is an absolutely essential part of the zoo," Gray commented. "The Friends can do what the zoo can't: raise money independently of the city." Gray cited five possible money sources: the city, the county, private donations, corporate donations and matching federal funds. "While $3.2 million is a large amount as a lump sum," Gray noted, "the renovation could be stretched out to five or six years." He acknowledged that the Friends had only about one hundred members and "had a hard struggle getting going," but he said the group "now seems to be stronger" and would coordinate the proposed fundraising.

The Audit Department of the City of Syracuse issued a report in 1975 that said the zoo director had submitted a capital development program of $3.3 million for the zoo for the years 1975 to 1980. "If enacted, the program would require $200,000 to be spent annually from 1975 through 1979 and $300,000 in 1980," the report stated. "The main building was constructed in 1936 and repairs and renovations are necessary. The City is unable to finance all of this program, but to date, the County has offered no financial assistance to the zoo." The audit report further noted that "because of financial problems, April 1, 1975 will see the institution of an admission fee to the zoo. Persons between 7 and 12 years of age will be 25 cents, while those aged 13 to 64 will be charged 50 cents. Others will be admitted free of charge."

The auditor said,

> *It is quite obvious that non-city residents make substantial use of city Parks and Recreation programs. Non-city residents participate in city recreation*

league sports events. They use city pools, ice rinks, parks and the zoo. All these facilities are funded solely by city taxpayers' dollars. If non-city residents wish to continue using these facilities, they must be willing to pay for the benefits they receive. Consolidation of city and county Parks and Recreation facilities would provide both city and non-city residents with the services they desire at a fair and equitable cost without the unnecessary cost of duplicate facilities.

The auditor concluded,

It is time for the City and the County to work together to officially provide the needed services to all our residents. It is an unnecessary luxury to maintain duplicate facilities such as the recent proposals to construct a new zoo and now a new library in the County. By working on a County-wide basis, we will have the necessary measures to more efficiently see to the needs of all our residents. City-County consolidation will come more easily in less expensive areas such as the zoo, parks and libraries. These areas will provide us with the opportunity to ease into consolidation, while at the same time they provide services which are in high demand by all residents of the County, both City and non-City.

The audit report's final words: "The key to successful consolidation in any area is firm moral and legal commitments by all parties involved."

Things came to a head when zoogoers realized that Siri the elephant, a favorite of visitors, had grown too large for her stall and was unable to pass through her thirty-five-inch-wide, seven-foot-tall cage door. "Syracuse just isn't big enough these days for the young lady who during the last year gained over 900 pounds and grew a foot," said a 1976 newspaper article. When Siri arrived in Syracuse in 1972, at age four and a half, she weighed 2,250 pounds. By 1975, she was nearly twice that weight at 4,200 pounds. In 1976, an elephant-sized key for an elephant-sized door was presented to Siri by Parke Wicks, president of First Trust and Deposit Company, and accepted on Siri's behalf by Linda Campbell, vice president of the Friends of the Burnet Park Zoo. The key, the official First Trust symbol, also denoted the bank's $200 contribution to the Friends fund drive to construct a new door to permit a fast-growing Siri to leave her enclosure. Despite the new doorway, which cost $2,000 to install, more help was needed for Siri and the rest of the zoo, which was "going downhill fast," according to John Gray, the zoo director.

In 1976, the county legislature failed to pass a bill providing funds to keep the zoo open. Zoo supporters warned that the lack of funding would result in the dismissal of six attendants, almost certainly forcing the zoo to close. The legislature's Ways and Means Committee tabled the motion to allocate money for the zoo because of confusion about how much money was needed to keep the zoo going for a year. The Friends said the zoo needed $207,000 to keep it operating that year, $40,000 of which was needed to pay for the six attendants whose federally subsidized salaries ended that year, but the motion introduced into the legislature asked for only $162,000, described as a "bare bones, status quo" allocation. The county budget director said he was told by the city that the zoo needed about $359,000 just for operating and maintenance costs. The motion was tabled until the figures could be checked.

A year later, Gregory Gualtieri, president of the Zoological Society, presented a plan for a $20 million zoo-recreation complex along the Seneca River to the Cayuga County Planning Board. Gualtieri said the plan would need a minimum of four hundred to five hundred acres but could cover up to one thousand. He added it was a "fantastic concept for the Auburn area" and predicted that it would bring large numbers of tourists to the area and attract industry. The proposed zoo would not have cages; rather, the animals would be separated by moats and other devices. Gualtieri was a man ahead of his time. He said a site on the river would provide water to fill the moats and wash down the animals. He added that the park, which he designed, would be built in self-contained sections, each costing about $2 million and taking about two years to construct. He claimed it would be a "self-supporting, moneymaking proposition" that would employ up to five hundred people.

County executive John Mulroy facilitated a partnership between the county and the city that allowed the zoo to enter a promising new era. *Courtesy of the Onondaga Historical Association.*

The debate over the zoo soon reached a boiling point. Syracuse's Mayor Lee Alexander and County Executive John Mulroy agreed that only the county could afford to pay for the radical improvements that were needed to keep the zoo operating. For two years, the zoo was jointly funded by the city and the county. None of the multiple proposals to address the situation, including moving the zoo to the shore of Onondaga Lake or the Seneca River, achieved consensus. A crisis point was reached when the Humane Society of Central New York issued a scathing report about conditions at the zoo, accusing the city of running an animal prison, crowding incompatible species in an "unsanitary crackerbox environment." Syracuse's Mayor Lee Alexander blamed and fired zoo director Charles Clift, although six years earlier, Clift had complained about the zoo's inadequate budget.

The City of Syracuse didn't have the money to make the improvements needed to bring the zoo up to standard, so Onondaga County stepped in. "The zoo is at the stage where we must do something or close it," said County Executive John Mulroy. On January 1, 1979, Onondaga County assumed full ownership of the zoo.

Chapter 3

A NEW BREED OF ZOO

Onondaga County's takeover of the Burnet Park Zoo provided an opportunity for change. A study by the County Department of Parks staff resulted in a forty-page renovation plan that involved shutting down the old zoo and constructing a new one. The plan was approved by the county legislature in 1981. The old zoo was closed in 1982, and the $12.8 million project began in 1983. Of the total cost, $10 million was provided by the county and federal grants and the rest was to be raised by the Friends of the Burnet Park Zoo. The new zoo was to be completely unlike its predecessor.

James Johst, the county's commissioner of parks and recreation, was not happy with the decision to take over the zoo. "I thought we were buying a pig in a poke," he said. But not one to shirk his responsibilities, Johst embraced the task. Most municipalities that built zoos hired an architectural firm and gave it an outline of what was desired. Johst did things differently. "There was a great deal of in-house talent both at the Parks Department and the zoo," he said. "There were some sharp people on the staff, and we felt we had a contribution to make." Johst created an ad hoc committee consisting of himself, Gary MacLachlan, zoo director David Raboy, Phil Suitors and John Eallonardo and charged them with developing a plan for a new, unique thematic zoo.

In a hard-hitting interview in the *Post Standard*, Johst was frank about the zoo's deficiencies and what he saw as its future. Asked to describe the conditions for animals at the zoo, Johst didn't mince words: "The zoo is absolutely atrocious. It's from years and years of being on the back burner

Elephant and county officials celebrated the reopening of the zoo.

and not being considered as important as everything else." Asked whether spending money to repair the zoo was worth it, he replied, "Any money that's spent ought to make it more habitable, more attractive, and offer visitors more information. But to patch up the plumbing, the water system and the heating system and leave everything else as it is would be a waste, in my judgment."

Although Johst acknowledged the county's desire to make things right at the zoo, he recognized the financial challenges involved. Nonetheless, he stated, "We have a responsibility if we are going to operate something to operate it well. We have a responsibility not to be slumlords." He went so far as to say that closing the zoo was "one option that is preferable to leaving the zoo in the condition that it's in right now." While praising the zoo staff for its dedication and commitment to the animals, he was extremely critical of the conditions in which the animals had to live. "If you ever go to the zoo, watch the jaguar," he said. "Watch him pace back and forth. I'm not an animal psychologist, but my estimate is that that animal is psychotic. It's living in a 10-by-10-foot cubicle." Still hopeful, Johst expressed confidence that the situation could be remediated if the master plan was approved and funded.

Despite differing priorities between recreation-oriented parks officials, whose goal was public entertainment, and zoo professionals, whose primary concern was animal welfare, consensus was achieved. Johst presented the county with a master plan for the overhaul. It was sent to the county's Conservation and Recreation Committee, chaired by Jack Haley. Legislators balked when they saw the estimated price tag of $10 million and ordered Johst to come back in a month with a scaled-down version. Haley, who had loved the zoo since he was a boy, angrily called Johst's plan "a Rolls Royce" when he first saw it, but he ultimately put aside his concerns about the cost and led the fight for the zoo in the legislature's Republican caucus. Legislative chair Nicholas Pirro insisted that one quarter of the funds be raised through a community fund drive. Finally, the legislature voted to bond $7.5 million for a new zoo if the Friends of the Burnet Park Zoo could first raise $2.5 million.

The Friends of the Burnet Park Zoo had been formed in 1970 with the goal of stimulating public interest in the growth and improvement of the zoo and assisting in the expansion of zoo facilities. Initially, the Friends consisted of a handful of volunteers trying to improve the then fifty-six-year-old zoo, but the organization worked hard to build up its membership. In May 1971, the Friends numbered 304. Three months later, following a mass mailing to prospective members and a raft of publicity about zoo enhancements, membership had increased to nearly 5,000. The organization held fundraisers to secure the money needed to enlarge the zoo to eighteen acres, build a fence to enclose the space, construct a boardwalk that traversed a portion of the property and establish a Western Plains habitat. Having helped to build the zoo, the Friends then worked to sustain it. The organization acquired animal specimens for the collection and planned to fund several capital improvements, including the purchase and installation of a floor-to-ceiling room divider for the community room.

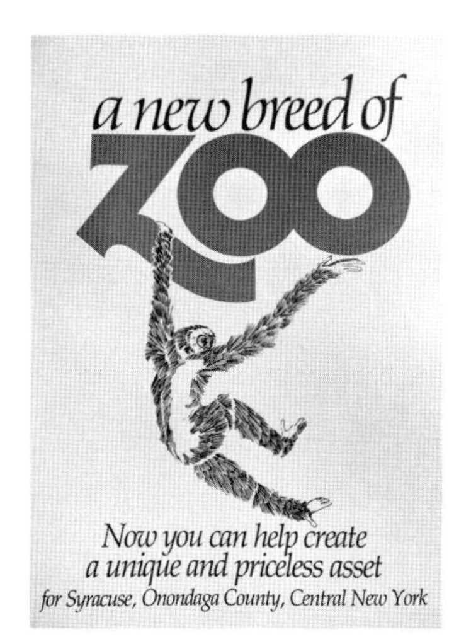

"A New Breed of Zoo" was the title of the zoo's campaign to create a unique and priceless asset, "for people, for animals, forever!" *Author's collection.*

"A New Breed of Zoo" was the title of the fundraising brochure created by the Friends of the Zoo Development Committee. "Few things are truly unique," declared the pamphlet. "Our new zoo is one of those few things. It will present a dramatic series of authentically constructed settings conceived to illustrate evolutionary, environmental and sociological aspects of all creatures and how they relate to human beings and the dangers we both face today."

The group, chaired by industrialist Raymond Cummings, engaged over 450 volunteers to raise the needed $2.5 million in corporate and private donations to supplement county and federal dollars. The effort was so successful that most of the money was collected within three months. Over three thousand donors contributed amounts ranging from twenty-five cents to $200,000. The Rosamond Gifford Foundation supported the project generously, as Ms. Gifford had been a lover of animals. Ultimately, the cost of the renovation was $12.8 million.

James Aiello, the zoo's director of education, explained that the new zoo was organized around three main themes, each subdivided into exhibit complexes to present concepts in an understandable form with species inventories, graphics, text and programs developed to enhance the thematic concepts. The three themes were Evolution of Life on Earth, Wild North and Animals and People. A full tour of the new zoo was estimated to take three to four hours, beginning at the statue of a Neanderthal man at the entrance to a three-hundred-foot-long dimly lit tunnel that housed the Animals in Antiquity exhibit, which displayed the first theme: the evolution of life on earth. A winding path filled with small glass tanks then traced the evolution of species from six hundred million years ago to the development of early mammals. Small blacktip sharks, rare turtles, shellfish, Cuban crocodiles and a tankful of Lake Ontario fish told the story.

The Animal Diversity exhibit focused on birds, emphasizing the way different species find food, avoid danger and raise young. Animal Adaptations demonstrated easily seen adaptations to specific environments (e.g., river otters to water, sloths to trees) and the role of adaptations in survival. Animals as Social Beings treated socialization as an adaptation for survival and exhibited animals with colonial or group social structures (e.g., lions, baboons, bees, weaver birds, hyrax). At the end of the Antiquity section was an aviary filled with exotic plants and many varieties of birds, chosen to illustrate the theme of Animal Diversity, as there are more than 8,700 bird species in the world. The zoo's selection included Bonaparte's gulls, meat eaters like the burrowing owl, fruit-eating toucans and cedar waxwings and colorful South American exotics like the scarlet ibis. Few of the birds were caged, so they flew past visitors as they walked through the exhibit. "The setting will be an all-glass environment filled with trees, lush plant life, pools and moving water—a setting alive with birds of all sizes, shapes and descriptions," the brochure enthused.

The twenty-six-acre Wild North exhibit was intended to explain the balance of nature as exhibited in four North American biomes: arctic tundra,

Thousands of kids enjoyed memorable rides atop the zoo's beloved pachyderm.

western highlands, eastern deciduous and western grasslands. Concepts included the fragility of environments, the importance of animals to overall balance, natural change in environments and the necessity and ability of animals to adapt to change in order to survive. Animals were exhibited in increasing order of human impact, including endangered (eagles, mountain lions) and extinct animals. The exhibit included a quiet place intended to induce visitors to think about what has been lost and the role of zoos in conservation. "Animal containment in this complex will be varied, imaginative and designed and built to minimize or eliminate any sense of containment," the brochure explained. It went on to say that "visitors will also discover some of the serious consequences for mankind inherent in our diminishing variant of animal life and some of the ways this critical situation is being resolved."

The Animals and People area included exhibits on animals benefiting people; the story of domestication; including artificial selection; the ways domestic animals serve people; and people benefiting animals (for example, how zoos provide for animals: diet preparation, veterinary care, nursery, etc.) "Children will particularly like this area," the brochure explained, "because they will be able to move among domestic animals, see a llama and, perhaps, take a ride on Siri, another elephant, or a pony. Everyone will enjoy the experience and become more aware of the importance of animals as providers of food, clothing, medicine, transportation, work and more for those human beings with whom they coexist." Visitors would move back indoors for this final exhibit complex. They would see and learn about the zoo's kitchen and the specialized diets prepared there, as well as the animal health clinic and nursery. Two-way communication between visitors and staff would enable the exhibit to both provide information and answer questions. "This will be a unique and memorable finale to any visit to our zoo," the Friends declared, "for the activities here will be ever-changing and ever fascinating."

David Raboy had been hired as zoo director one month before the county took over. In 1982, Commissioner Johst announced that the zoo would be closed for two and a half years. The animals were "to be sold, shipped elsewhere or just put out to pasture when construction begins," Raboy explained. Zoo favorite Siri was to be sent to the Buffalo Zoo. "We hope that either she comes back pregnant or we will purchase a second elephant," Raboy said.

Animal keeper Charles "Chuck" Doyle, who had begun working at the zoo in 1976, had developed a particular bond with Siri, the strength of which was tested when it was decided that Siri would go to Buffalo. Although happily married, Doyle announced, "Let me put it like this. Siri's going to Buffalo. My wife is staying in Syracuse. I had my choice. I'm going to Buffalo."

Animal keeper Charles "Chuck" Doyle developed a special bond with Siri, which led to a long career teaching others about elephant care and training.

Siri's relationship with Doyle was unique. She was affectionate with him and listened and responded to him. She knew and followed over thirty direct commands. Zoo director Raboy noted that Siri obeyed Doyle because she wanted to. "You can't manhandle an elephant," he pointed out.

Buffalo's elephant, Lulu, had a keeper but no handler and nothing like the training Siri had. The Buffalo Zoo was eager to have Doyle's expertise. Zoo director Minot Ortolani said, "We feel that with a handler here, our elephant will learn a lot." Erie County did not charge for Siri's stay, which cost about $21,000 per year, and subsidized 75 percent of Doyle's rent, thereby obviating the complaint of one Onondaga County legislator that the county was paying Doyle to "babysit an elephant." Siri was a big hit in Buffalo and, after a slow start, got along well with Lulu.

Doyle was promoted to senior keeper in 1982 and zoo director in 2006. He was committed to the welfare of the animals and to close relationships with the staff. His approach to elephant care laid the groundwork for the zoo's Asian elephant exhibit and successful breeding program. Doyle became a consultant for other animal programs, served as executive director of the Elephant Managers Association and taught a course in elephant training principles for the American Zoo and Aquarium Association. "I don't know anybody who isn't fascinated by elephants," Doyle said in an interview in the *Syracuse New Times* in June 2011. "Whether it's their size or their unique nose, their intelligence—everything. Animal behavior is a fascinating subject, no matter what the species."

Doyle sought to maintain a balance between the entertainment aspect of the zoo and the desire to put animal welfare and education first. He said that visitors should definitely "have fun while you're here," but he also wanted to give them "an understanding of the relationship between animals and people and action to sustain the environment we share." He elaborated,

> *If you care about elephants or you care about wood ducks, then you're going to care about the environment and do your small part to protect it. I don't care if a kid comes here, listens to the elephant demonstration and doesn't know how many muscles are in an elephant's trunk or how long they can live. That information is gravy. But if they come away and care that the elephants are here and in the wild, then we've met our mission. Hopefully we can provide them some means to act on that.*

He added, "We're not going to be able to get anyone to pay attention if we're not a little entertaining. The zoo is a fun place; it's got to be a fun place.

But it doesn't have to be exclusively fun. There is an educational message, conservation message in everything we do."

Siri was so gentle and well-trained that she gave between eight thousand and nine thousand rides a year to children. Doyle cowrote an article with Donald Moore titled "Elephant Training and Ride Operations, Part I: Animal Health, Cost/Benefit and Philosophy," published in the journal *Elephant* in 1986. In the article, the authors shared the results of a survey they conducted as employees of the Burnet Park Zoo that showed that very few captive elephants in zoos (eighteen in the United States) were trained for ride operations. They contended that trained elephants are "easily accessible for treatment, are less 'bored' and overall are healthier than non-trained elephants, which may be manifested in a longer life span." Doyle and Moore argued that "the benefits derived from a well-planned elephant training and ride operation clearly outweigh the costs incurred." Their article asserted that "the importance of training elephants cannot be underestimated: it allows personnel to meet management needs, reduces inactivity, and may be important for monitoring animal health. In addition, trained 'ride elephants' may return more in other revenues and capital funding than is earned simply from elephant ride ticket sales."

In 1985, Siri returned to Syracuse. The zoo acquired three other elephants: eight-year-old Romani, twenty-eight-year-old Babe and thirteen-year-old Indie. The quartet moved into the zoo's new elephant management facility. A baby elephant, born in 2009 to Mali, one of the zoo's Asian elephants, who was in Canada awaiting return to Syracuse, was named Little Chuck to honor Doyle. Although zoo officials referred to the elephant as Little Chuck, he wasn't so little. He weighed 235 pounds at birth.

On August 2, 1986, the new Burnet Park Zoo officially opened its doors to the public, with baby Romani breaking the ceremonial ribbon. Crowds of up to twelve thousand came to inspect the new facilities. "The zoo contained more than 300 species and over 1,000 specimens, making it three times the size of the old zoo and was second in size to the Bronx Zoo among New York's zoos." Jettisoning the old model of animals in cages, "the zoo featured whole animal social groups in natural settings, and illustrated the evolution of species and their relationships with each other, the environment, and man." During the dedication ceremonies, public officials praised the zoo's reopening. Syracuse mayor Tom Young called it "a very, very special day," and county legislative chair Nicholas Pirro, commending the zoo's staff, said, "I am sure they will make the zoo second to none."[9]

The Friends of the Zoo were very proud of the role they played in the creation of the new zoo. Thousands of Friends members came to a preview of the zoo prior to its official opening. Anita Monsees, a cofounder of the Friends, said, "I think it's a chic thing to do right now and I really think that's why most people are joining, but I hope that as a result they will become much more interested in preservation of animals. This is such a fabulous zoo. It's the kind of zoo you never absorb in one visit."

The *Post Standard* was still harsh in its report of the new zoo.

> *The new Burnet Park zoo was born of communitywide dissatisfaction with the crumbling, unsanitary old zoo....In the memory of most Syracusans, a trip to the Burnet Park Zoo meant walking into a smelly, noisy building that housed animals in rows of cramped cages....What has grown from the rubble of one of the country's worst zoos is a facility that is the first in the country to use evolutionary science as its primary theme.*

The paper went on to state,

> *The contrasts to the old zoo are everywhere. The small enclosure where Siri the elephant was kept indoors for nearly two years because she had outgrown the door has been replaced by a large exercise yard and a spacious elephant house with rubber flooring to prevent its inhabitants from getting arthritis. Where once bears and lions were kept in stinky cages barely large enough for them to turn around, there now stands in airy building with exhibits specially designed to simulate the animals' natural habitats. Outdoors some*

The opening drew a crowd of over twelve thousand people for the weekend. The final cost of the renovation was $12.8 million, and it took more than seven years to complete. *Courtesy of the Onondaga Historical Association.*

of the habitats are so natural it is difficult to find the animals. Only patient visitors will get a glimpse of the red wolves, for example, who live on a thickly wooded lot in a far corner of the zoo.

The paper devoted several pages to stories of the zoo's deficiencies in past decades, in a special section entitled "Rising from the Rubble." It seemed unable to forget the zoo's past, even in articles about the new concepts and methodologies that characterized the new zoo. The paper kept reminding readers that "cramped quarters, poor sanitation, slipshod veterinary care and exhibits that bore no resemblance to natural habitat" had earned the zoo the rating of "one of the ten worst in the nation."

The negativity of the press notwithstanding, the renovated zoo was a state-of-the-art facility for its time and was considered to be one of a "new breed" of zoos. It did more than just display animals. Instead, it also functioned as a learning center, demonstrating how animals evolved, adapted and eventually became endangered. "It's like night and day," said Robert Geraci, Onondaga County parks commissioner. "We're talking comparing your basic jail cell with cement block walls, iron bars and one tenth the space

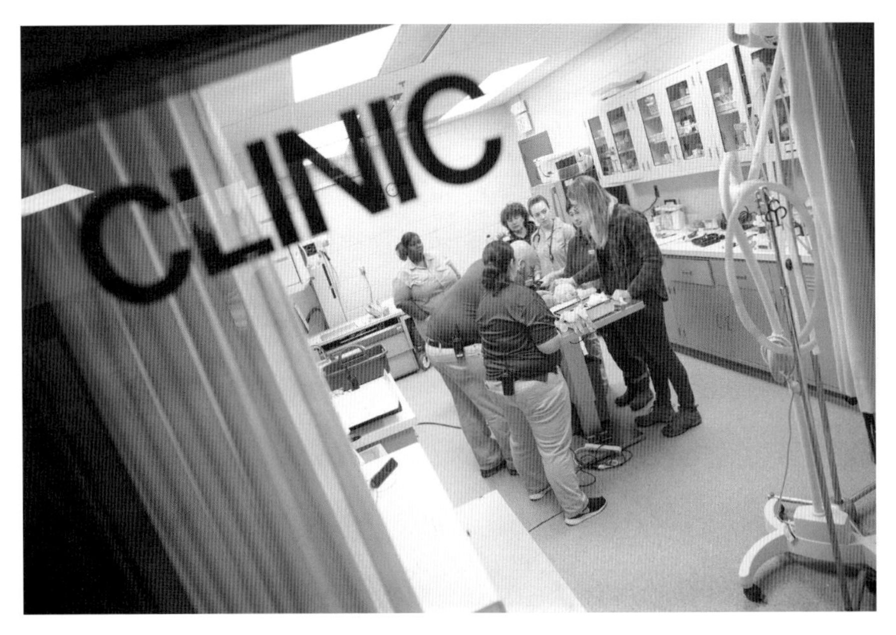

The new veterinary clinic allowed all animal treatment to be done on-site, utilizing a fully equipped surgical room and an operating table.

that animals needed to live comfortably, to very roomy exhibits that put the animals in their natural habitats and are aesthetically pleasing to visitors." The reaction of zoogoers was overwhelmingly positive. "Fantastic," "Love it, it's a big difference," "Worth the price," "A zoo we can be proud of," were some of the comments recorded in the local paper.

Within a year of reopening, the zoo received its first accreditation from the Association of Zoos & Aquariums, certifying that it met the highest standards of animal care and welfare, wildlife conservation education and guest experience. One of the most valued additions was the veterinary clinic, which the old zoo lacked. Previously, all animal treatment had to be done off-site. The new clinic was equipped with microscopes, centrifuges for analyzing blood samples, a portable X-ray machine, a fully equipped surgical room, an anesthesia machine, tracheal tubes to fit the throat of any animal and an operating table. The clinic featured a picture window looking in from the main hallway so that visitors could watch veterinarian Lisa Jensen treat her patients. "It should be a real educational experience," said Jensen, "although I don't think people will want to see everything that's going on."

On the zoo's one-year anniversary, the *Herald Journal* waxed enthusiastic:

> *The new-look zoo has seen success in its first year, success beyond the wildest imagination of everyone who worked to make the changes happen. Zoo officials expect they'll have had 600,000 visitors by the end of 1987, almost 200,000 more than the most optimistic predictions before the zoo opened. And the success is reflected in the zoo coffers. Revenues total more than $750,000. There were 300 friends of the zoo; now, the zoo has more than 10,000.*

The newspaper went on to rave,

> *Those who remember the old zoo (1914–1983) stroll the grounds these days as if something magical happened there. They see what was once a smelly old eyesore with a menagerie of animals is now a beautiful place with flowers and fountains, wooden walkways and bridges, where roughly 1,000 animals representing more than 350 species live in environments that simulate their homes in the wild.*

The 100th anniversary celebration was quite the event. It was kicked off with 10,000 kazoos playing a musical salute, followed by a giant birthday cake big enough to feed 5,000 people 5,500 animal crackers. An airplane flew through the sky with a banner wishing a happy birthday to the zoo, and a special birthday cake was prepared for Siri, who was turning 19. The new zoo was "a magical place," the newspaper declared, and a recent visit by television personality Captain Kangaroo seemed to confirm the opinion that things were moving in the right direction.

Elephants continued to feature prominently in the zoo's success. An Asian Elephant Celebration was an annual event. A 1989 brochure for the event featured statistics about the zoo's four pachyderms, along with a list of "ele-facts" and a tally of their annual food consumption: 172,800 pounds of hay, 13,140 pounds of grain, 12,125 pounds of carrots, 274 bushels of apples, 185 bushels of yams, 185 bushels of bananas, 4,360 oranges and 58,000 gallons of water. The brochure also emphasized that "Asian elephants are an endangered species. They are being forced out of their natural forest habitat and, unfortunately, many males have fallen victim to poachers because of their giant ivory tusks."

ELEPHANT CELEBRATION
AUGUST 19-20, 1989

Welcome to our annual Asian Elephant Celebration.

Cultural and religious significance has been given to elephants for centuries. Elephants are treated as "gods" in Asian cultures, but not in a western sense. The elephant "god" is a sign of great strength.

Asian elephants are an endangered species.They are being forced out of their natural forest habitat and, unfortunately, many males have fallen victim to poachers because of their giant ivory tusks. The increase in truck-logging has also led to a decline in the use and production of domesticated work elephants in the hardwood forests of Asia. Less than 40,000 Asian elephants survive in the entire world. Zoos are trying to breed Asian elephants in captivity to preserve these magnificent animals for future generations.

The birth of Tundi in 1991, the first successful elephant birth at the zoo, was cause for celebration and resulted in large numbers of visitors. *Author's collection.*

On July 10, 1991, elephants again made the news. The Syracuse *Herald-Journal* ran a front-page headline that trumpeted, "It's a boy!" A 274-pound calf, officially named Tundi, meaning "beloved one," had been born to Romani—the first successful elephant birth at the zoo. The event was significant not only in its own right but also as a way to overcome the profound grief felt the year before when a baby elephant carried to term by Babe was stillborn and Babe died as a result of complications from the delivery. Romani's calf was welcomed by jubilant and record-breaking crowds, who called the tiny elephant Emmett before his official name was bestowed.

Zoo director David Raboy discussed Tundi's arrival in the fall issue of *Tracks*:

> *The successful birth, while viewed as a single event by the public, is actually the result of a multi-faceted elephant management program and should be viewed as one step within a larger comprehensive program. In this light, our management of elephants parallels very closely our management of most*

On July 10, 1991, proud pachyderm parents Romani and Indy announced the birth of their bull calf Tundi, weighing in at 274 pounds and growing fast.

> *other species here at the zoo. Unfortunately, many births rarely draw the public attention that an elephant birth produces. These less dramatic births do not become "events" even though the effort, training and dedication of the staff is just as strong as it is among our elephant keepers.*

Raboy pointed out,

> *Creating the essentials for the captive propagation of elephants is indeed difficult. But if you stop for a moment to consider management and breeding successes of animals such as amphibians, like the poison arrow frogs, or birds, like the hyacinth macaw, or, for that matter, several of the more unusual small mammals such as the Hoffmans sloths or ruffed lemurs, you will quickly come to the conclusion that these species also present difficult challenges to be overcome before we can count our efforts as successful. I'm always amazed at the skills that staff exhibit and the successful propagation of such unusual animals as the panther chameleons whose eggs must be carefully monitored and nurtured throughout the incubation and hatching process.*

Raboy concluded, "I'm sure my point is now obvious: while an elephant birth is a wonderful experience with much attendant publicity, we in the zoo field and you, our strongest supporters, must never lose sight of all the really fine accomplishments of the entire staff throughout the zoo."

In October 1991, the zoo hosted the Twelfth International Elephant Workshop, a four-day series of presentations by zoo officials from Bangalore, India; Syracuse; Jacksonville; Toronto; Honolulu; Oakland; Los Angeles; Milwaukee; Phoenix; Atlanta; Seattle; San Diego; Washington, D.C.; and other cities. There were also programs led by experts from the SUNY Health Science Center and Cornell University. Friends of the Burnet Park Zoo hosted meals, and Romani and Tundi gave a demonstration of their training. Topics included "Management of Captive Elephants," "Tusk Extraction in an Unmanageable African Bull Elephant," "The Birth and Attempt at Hand Raising an Asian Elephant Calf" and "Do Elephants Really Forget?"

But the zoo was still evolving, and six years later, director John Gray drew up a master plan. His statement of purpose was strong and forthright and deserves to be read in its entirety.

> *The time has come to reorganize the goals and design of our zoo. This should not be just a face lifting, but a thorough introspection of the reasons behind our having a zoo. It is no longer justifiable to remove animals from the wild to place them on public display for recreation. With the vast coverage wild animals receive on television and in the popular press, it is no longer necessary for a community to show caged animals in a city park. But, there is a need for facilities and programs to teach the public about wild animal neighbors, their role in our lives, and how our lives affect them.*
>
> *This community does not have a natural history museum. The present zoo is merely a random collection of animals with no purpose. I propose that a new, all-weather facility be built in Burnet Park, to incorporate the present zoo where practical. This new facility would combine the best features of a natural history museum and zoo and would therefore be more aptly named the Onondaga Zoological Park or the Onondaga Natural Science Center. The responsibility of this center would be to display animals and other biological material, both living and static, in an educational and enlightening manner.*
>
> *The facility will be designed to encourage maximum use by both organized educational groups and the citizens at large. Research and conservation goals will also be met, but the primary function of even these will be for public education. This plan does not depict an extravagant*

> *facility and it should not be considered superfluous. I feel that the plan I have set forth is the minimum necessary to optimize our investment in the education of the children and voters of this area in the biological sciences, conservation, and ecology.*
>
> *Ideally, the entire program would be funded and built at one time. Recognizing the financial problems of the economy today, an orderly progression of facility construction may be adopted, providing a commitment for completion is made. (Each facility in Section II is annotated with a priority number and placed in order of construction.) Funding for construction must come from every possible source: City and County capitol funds, State Education funds, Federal Education and Conservation funds, local foundations, educational institutions, public subscriptions, and entrance charges. Since the educational wealth of the improved facility will be shared by the entire community, the operating budget of the new facility should also be shared by the entire community: City, County, and State. This assistance with the operational budget can take many forms: from direct cash payment to the appointment of teachers operating under school budgets to manage the educational program.*
>
> *The actual design of facilities and the completion of the plot layout will wait until this basic proposal for development is acted upon. All of the buildings herein described will be of a radically new design for zoos. Yet, they will represent the lowest cost buildings in modern zoo construction. Individual displays will not be built during a building's construction. Only the shell, with attending utilities, will be permanent. The individual displays will be mobile modular units. The floor plan will be flexible and exhibits will be changed as new approaches or problems are found. The buildings cannot become outdated because their interiors will change with the times.*

Gray went on to outline the various buildings proposed in his plan, in priority order. First and foremost was the Education Building. "This one building should be considered the most important structure in the zoo," Gray said. "It will house the education division and all of its activities. Since every feature is designed for multiple use, it will be fully utilized year-round." The Education Building was to house at least three classrooms and three laboratories, a library, zoo and education offices, a graphics preparation room, a planetarium and restrooms. A sister building was envisioned as the Animal Kingdom Building, which would "display the smaller animals in a living museum setting."

The second priority was a visitor center, which was "intended to introduce the visitor to our facilities and intrigue him with the educational aspects of the zoo." It was to include an exhibit complex and displays. "All walks would be one way to ease traffic and confusion and to enable the exhibits to be presented in a more meaningful sequence," Gray explained.

Priority 3 was the Ecology Building, which "would present ecology in a manner to better inform our citizens." It would explain the role of man in the world around him through multiple displays and animal exhibits and would include a kitchen, veterinary and quarantine rooms and a baby animal nursery. "We are all responsible for our environment and the legislation to protect it," Gray wrote. An upgrade of existing facilities was included in this priority. The role of the Ecology Building was broadly defined.

> *This building would explain the role of man in the world around him. It would define ecology in terms and graphics understandable to the broadest audience. It would detail the short- and long-range plant problems of pollution, pesticides, over hunting, and lack of hunting, introduction of exotic animals, mechanical destruction of environment, etcetera. It would describe the complexities of a food chain and put it in terms of energy at each level. It would show a normal beneficial chain and a detrimental chain (vermin, overgrazing, etc.) with representative species. It would explain natural population controls and population cyclic* [phenomena]. *It would demonstrate range, habitat, niche, territory, individual distance, migration, dispersal, etc.*

Priority 3 also included a service building. "Facilities presently available for service would be totally inadequate for the additional facilities proposed," wrote Gray. "At present some of these service facilities are severely taxed or nonexistent. The public finds these service programs most interesting and presently misses a great deal by not having access to them. For example, many baby animals are born, raised and sold without their ever having been seen by the public, due to a lack of exhibition facilities." Gray proposed the addition of a kitchen, a walk-in cooler, a freezer, dry storage and an automatic recording weather station visible to the public through a picture window, as well as a nursery and quarantine and veterinary sections. He also proposed that certain existing areas be brought up to the standards of the new facilities. These included the feline building and primate island, the bird areas, ponds and enclosures and the hoofstock yards and barns. "Every exhibit would be well explained through appropriate graphics: comparison of horns and antlers; how they are used in combat; what is the mechanism whereby a deer can grow and lose his antlers yearly."

Priority 4 was a research facility, primarily focused on animals in the wild. "All of the schools and research facilities in this area are interested in behavior studies," Gray explained. "Most of them have behavior study rooms or labs, but all lack the large areas that our zoo could provide. Due to space limitations, most research is now being carried out with the domestic forms of animals, while the real need is to study the wild forms." In addition to laboratories, Gray envisioned two enclosed "flight pens" and a "public interpretation room." He saw this building as "a fitting climax to our building program," which would enable the zoo "to do our part in the breeding of an endangered form of life for eventual re-stocking in the wild."

It is not clear whether Gray's plan was ever given serious consideration.

In 1997, the *Post-Standard* declared that the zoo was "a whole new animal" and wished it a happy tenth anniversary, wiping out almost a century of past history as if it were a bad dream.

> *Long after many zoos had begun building exhibits that attempted to recreate the habitats of the animals in their collections, giving the animals more space and more stimulation, and the people visiting them a sense of how animals really lived, the old Burnet Park zoo still housed its animals in what Doyle calls cell-block style. Iron mesh and concrete cages containing board and listless animals.*

It continued, "The new zoo differs from the old zoo in more than physical design. The new zoo has a commitment to education, science, conservation and preservation of animals that the old zoo lacked."

The *Post-Standard* devoted a lot of space to a discussion of elephants in its birthday edition. A front-page article entitled "The Tracks of My Tears" explained that elephants weep not because of melancholy but to keep their eyes clean. A lot of column space was devoted to the question of where five-year-old Tundi would be next year and quoted curator Doyle's observation that Tundi "was starting to be sexually interested. He's enthralled with his little sister." The paper noted that in the wild, adult males roam alone and again quoted Doyle: "He has to be separated for the herd's safety and his safety." Doyle hoped that Tundi would be able to go where he could grow to maturity with a group of young females and start a breeding herd.

Chapter 4

ROSAMOND GIFFORD AND GROWING FROM STRENGTH TO STRENGTH

Rosamond Gifford was born in 1873 to William H. Gifford and his wife, Mary Augusta Skinner. Gifford was a well-known attorney in Syracuse and a former district attorney. Rosamond was sent to boarding school in Wellesley, Massachusetts. There she met an older man named Fred LaFayette, a railroad gambler by profession. Rosamond and LaFayette lived together for several years and married in 1895. The marriage lasted five years, and after the divorce, Rosamond took back her maiden name. She then moved to Boston to study the harp. In 1913, her father convinced her to give up her musical career in exchange for his entire estate, contingent on her agreement to live on and care for her father's farm on Thompson Road for the balance of his life.

William Gifford died in 1917, leaving an estate of over $1 million. Rosamond was burdened by the inheritance. An article in the November 23, 1920 *Syracuse Herald* quoted her as saying,

> *Do you know that since Dad died, I have probably had 1,000 or 2,000 letters and that practically all of them have asked for money? These letters are practically all begging letters. They ask for all kinds of sums from $.35 to $100,000. I could have given away every penny that my father left twice over, if I had responded to even a quarter of the requests I had received. I have been asked to send boys and girls through college. I have been asked to endow hospitals. I have been asked to build churches of every denomination and to found scholarships and take over business concerns. I have been invited to invest in at least 500 get rich security and stock companies.*

Above: Rosamond Gifford was an heiress who left her wealth for "religious, educational, scientific, charitable or benevolent uses." *Courtesy of the Onondaga Historical Association.*

Right: Rosamond Gifford wanted to help people but said, in an article in the November 23, 1920 *Syracuse Herald*, "I haven't decided yet just how it ought to be done, but I don't think it will be through organized charity." *Courtesy of the Onondaga Historical Association.*

Despite having inherited so much money, Rosamond lived an isolated and spartan life. One news report from the time recounts that her standard dress for meetings with her lawyer and banker was a leopard skin coat and riding boots. The large home she lived in was sparsely appointed, and her bedroom was the only furnished room on the second floor. When she died in 1953, her executors had to borrow chairs to hold the very small funeral in her home. The barn contained thirty-three goats, whose milk was used to feed her more than fifty cats and Rosamond herself, who was a great proponent of goat's milk.

Onondaga Historical Association director Gregg Tripoli wrote,

> *Rosamond was a tough, no-nonsense woman, there's no doubt about it. She fought a few high-profile court battles in her lifetime over money and she was just as protective of her nest egg as her father was about his. By the time of her death, she turned the $1,000,000 that her father took a lifetime to build into an estate worth $6,000,000 (the equivalent of just over $48,000,000 today) and she gave almost every penny of it to establish a charitable corporation.*

Rosamond was determined to be philanthropic. "Please don't think that because I have no idea of handing over my money to everyone who asks for it that I don't intend to do any good with it, or that I mean to hoard it," she declared. "I do want to help people. I haven't decided yet just how it ought to be done, but I don't think it will be through organized charity. I shall take the advice of people I know I can trust and then I'll try to get at those who need help and are too proud to ask for it, if that seems to be the best way." Rosamond died on April 15, 1953. Ironically, she died on the day federal taxes are due, as she annually sent payments to what she referred to as the "Infernal Revenue Service." She left the bulk of her estate to an organization formed exclusively for "religious, educational, scientific, charitable or benevolent uses known as the Rosamond Gifford Charitable Corporation."

The Rosamond Gifford Charitable Corporation was formally incorporated on July 23, 1954. The corporation started with approximately $5 million. Rosamond had invested well and had spent little of the funds she inherited from her father. At one time, she was the largest single shareholder of AT&T stock. Until the late 1990s, the corporation's board was composed almost exclusively of members and descendants of its founding families and their close business associates. The last of the original trustees, daughter of

Ms. Gifford's investment advisor, died in 1991. Eight years later, a newly invigorated board took a leadership role to help the community to identify needs, prioritize projects and apply Gifford resources. In its own words, the Gifford Foundation "supports individuals and organizations through grants and initiatives that build on community assets and promote positive change in the community." The foundation is "dedicated to the stewardship of the funds entrusted to its care" and "committed to using its financial and human resources to build the capacity of individuals and organizations to enhance the quality of life for the community."

"Through initiatives, grantmaking, and community engagement, the Gifford Foundation directs its support intentionally to meet organizations and individuals where they are and assist them in attaining their stated goals through capacity building." In 1999, the Rosamond Gifford Charitable Corporation voted to donate $2 million to the Burnet Park Zoo, with the proviso that it be renamed the Rosamond Gifford Zoo at Burnet Park. The zoo's mission statement read: "The mission of the Rosamond Gifford Zoo at Burnet Park is to conserve, exhibit and interpret a living animal collection in order to promote public recreation."

The Gifford Foundation's $2 million grant was a turning point in the zoo's evolution into a modern, first-class institution.

The Rosamond Gifford Foundation's $2 million endowment gift was a godsend. The zoo entered an era of accomplishment. It became the second zoo in the nation to successfully raise red panda triplets. Asian elephants, Amur tigers, Humboldt penguins, snow leopards, golden lion tamarins, markhor goats and red pandas, all endangered species, were bred successfully.

Dr. Anne Baker became zoo director in 1993 with a specific mission. "I really felt there was a lot of potential for a smaller zoo in a smaller community to have an impact in conservation education since the zoo was so much of the focus," she said. "It was an opportunity to get people to think more about the natural world, how they impact it and how they can lessen that impact." Baker had begun her zoo career as a predoctoral fellow at the National Zoo in Washington, D.C., studying the development of social behavior in primates in Sri Lanka. After receiving her PhD, she became curator of primates at the Brookfield Zoo, just outside Chicago.

When she moved to Syracuse to head the Burnet Park Zoo, Baker worked hard to make the zoo better for both animals and people. "We tried to expand the revenue-generating potential of the zoo, to allow us to do more for conservation," she commented. One of the first projects she initiated was renovating the zoo's front entrance area, making it the first green building in Central New York. "When I got here, the entrance didn't allow for good visitor traffic, and we needed more space to gather up school groups," Baker said. "We created a much better entrance with gardens outside containing native species and a flowing pond collecting water from the roof. We also expanded the gift shop and created a facility available for rentals to generate money."

"We did our own catering, which really helped, as that brought in quite a bit of revenue," Baker remarked. "We started an event called Brew at the Zoo and set up brew stations around the zoo after hours to get adults to come in. That was a good way to expand our audience to young adults who might not come to the zoo otherwise." The program proved to be extremely popular and became an annual adults-only event that attracted over two thousand visitors each year.

Opposite: Dr. Anne Baker served as director of the Rosamond Gifford Zoo in Syracuse from 1993 to 2006 and was valued for her leadership and commitment to animal science.

This page, top: The zoo's front entrance area was renovated to make it the first green building in Central New York and improve visitor access and traffic flow.

This page, bottom: Brew at the Zoo quickly became a popular summertime fundraiser, featuring an array of beer and wine stations, food trucks and live music.

Onondaga County executives, like Nick Pirro (*pictured*), have always been very involved in the zoo, even coming face to face with its elephants.

The zoo also worked to strengthen the quantity and quality of its animal habitats. "We renovated the primate area and made it into more of a rainforest theme," Baker recalled. "We did a submarine exhibit that took people back in time and underwater through the cycle of evolution. We thought, *If we want people to come to the zoo, we should have animals they want to see.* We did a survey to see what animals people would like to see, and tigers and penguins were among the top five." Baker realized that "both tigers and penguins were part of collaborative breeding programs, and the Humboldt penguin Species Survival Plan was looking for more institutions to have Humboldt penguins. We had a fair amount of space for tigers and gave them two really nice outdoor habitats. The penguin exhibit had to fit into a certain space, but it contained enough space to work and create an immersive experience." A habitat for Amur tigers was added to the zoo's main loop in 1999, the Niagara Mohawk Rainforest exhibit opened in the Social Animal Building in 2000 and the Humboldt penguin exhibit opened in 2005.

Baker helped create a stronger focus on animal welfare and conservation. "We highlighted animal welfare as a very important aspect of what we did at the zoo," she elaborated. "It was at the top of mind at our insistence. You want animals to have good welfare." The zoo put considerable efforts into saving animals in New York State. "We worked with the Department of Environmental Conservation on amber snails (a highly endangered species) and pond turtles," Baker said. "We focused on local species. Since there wasn't a lot of money for conservation, we thought it was most important to concentrate on projects in the community."

In 2005, the Rosamond Gifford Zoo made television news as it investigated becoming the first zoo in the nation to be powered by its own animal waste—in particular, "the prodigious piles produced by its pachyderms." The zoo, described as "world prominent for its Asian elephant breeding program," was investigating the feasibility of using animal waste as an alternative energy source to reduce its $400,000 annual heating and electricity bill. The zoo's six elephants produced more than one thousand pounds of dung per day, according to director Baker, who explained that while zoos are about conservation and stemming the loss of animals and habitat, "conservation is also about how people use natural resources. This is an opportunity to give visitors the whole picture."

Baker became a leader in the zoo field through her involvement with the Association of Zoos & Aquariums (AZA). She served as the organization's board chair for a year while at the Rosamond Gifford Zoo. She was the first female chair of the organization in fifty years. "That was a very interesting time," she reflected in an article on Zoophoria.net:

> *Elephants in zoos were a big topic in the media when I was president. I went to the AZA Animal Welfare Committee and said, "I want to know what makes a happy elephant."* Happy *is obviously an anthropomorphic term I used loosely, but I wanted to know how we would know if an elephant had good welfare. That was the kernel that got the whole study of elephant welfare started. If you look at new elephant exhibits that have been developed since, they've really relied on that study and paid attention to its findings.*

The Friends of the Zoo, which had begun as a small group of dedicated volunteer supporters, underwent a major transformation around this time, becoming a key partner in supporting the zoo's operations. Originally, Friends members trained to become docents, leading school and organization groups on tours, and worked in the zoo's gift shop. The shop featured "high-quality stuff, most of it with an educational bent," and also carried "things that schoolkids can get with a quarter." An early member of the Friends of the

Zoo recalled those early days: "There were a half dozen of us who always went to the zoo and thought, 'Gee, this is awful. It should be a lot better than this.' So we started Friends of the Zoo, and of course, we had no money, just a lot of goodwill. We raised money in the usual kinds of ways, and we stumbled along and limped along for a long, long time."

In 1971, encouraged by zoo director John Gray, the group began a membership drive, offering free admission to the zoo to Friends. The group grew to 400 members and then to about 700 by the late 1970s, when it played a role in supporting the county takeover of the zoo. When the zoo closed for renovations in 1982, Friends membership dwindled to 300, but another extensive membership drive was undertaken, offering an array of benefits. In addition to free admission to the zoo, those who became members received a subscription to *Tracks*, a monthly newsletter that outlined new activities and profiled animals and curators; a 10 percent discount in the gift shop; and invitations to special events. Membership categories and donation levels ranged from senior citizens and families at $15 and $30 to supporting and sustaining memberships at $100 and $250, respectively. The result was a robust membership roster of 6,800. When the zoo reopened in 1986, the role of the Friends changed considerably. Its membership fees were used to finance special activities, such as sending curators to seminars, training docents and offering educational programs.

The zoo was still not exempt from tragedy. In 2005, Kedar, a male calf, was born to twenty-two-year-old Targa, and his arrival was greeted with joy. His name meant "powerful" or "mighty one" in Hindi. But four days after his birth, Kedar fell into the pool in the elephant exhibit. When the other elephants went to help him, they mistakenly pushed him deeper, and he got water in his lungs. Despite being put on antibiotics, the little elephant did not survive. Inspectors said the pool should have had a barrier. Director Baker responded that the exhibit has been unchanged since 1991 and the Department of Agriculture had inspected it every year since then without raising concerns. The federal Department of Agriculture ordered the Rosamond Gifford Zoo to pay more than $10,000 in penalties for improperly handling a baby elephant.

Janet Agostini became head of the Friends in 2006 and turned it from a deficit to a profit center for the zoo, raising over $30 million during her tenure.

The role of the Friends kept expanding, and the organization was increasingly tasked with major fundraising. In 2006, Janet Agostini was hired as the Friends' executive director. At the time, the organization was $2 million in debt. Agostini worked closely with Friends staff and its board to transform the organization into a financially healthy partner for the zoo. Under her leadership, the organization became debt-free and consistently ended each year with an operating surplus. Agostini raised more than $30 million in earned and unearned revenue to support zoo operations, conducted several successful capital campaigns and oversaw the expansion of both the zoo's banquet facilities and the gift shop.

Lion cubs Kierha, Mindine and Joshua were born on April 21, 2000, at the Baton Rouge Zoo and came to Syracuse as eight-month-olds. According to the *Syracuse Post-Standard*, they weighed ninety pounds apiece and arrived in Syracuse aboard a specially equipped truck that kept temperatures at sixty degrees Fahrenheit. "On the first day of their release into the exhibit, an intense snowstorm arrived, coating every inch of the lions' home. While 9-month-old Joshua and Mindine braved the elements for a short frolic in the snow, it was too much for their shy sister, Kierha. She stayed indoors."

The public enjoyed watching the triplets grow. Kierha outlived her sister and her brothers, who passed away in 2014 and 2015, respectively. In 2015, the zoo acquired male lion M'wasi from the Bronx Zoo as a companion for Kierha. The two became close and could often be seen cuddled up together on exhibit. Kierha passed away at the age of eighteen; M'wasi died the same year.

Lions Kierha and M'wasi formed a special bond and were often seen cuddled together in their exhibit space.

The penguin exhibit opened to great acclaim in 2005 with six viewing stations to watch the nineteen penguins swimming and waddling.

In 2001, construction began on a Humboldt penguin exhibit. Humboldt penguins come from the Humboldt Current, a cold, low-salinity ocean current that flows north along the western coast of South America off the coast of Peru and Chile. Since they are from a temperate climate, they can tolerate Syracuse winters as well as summer heat. Their large, deep and complex pool is heated in winter and cooled in summer to keep the water at their preferred temperature of fifty to sixty degrees Fahrenheit year-round.

The penguin exhibit opened to great acclaim in 2005 with nineteen birds and has been recognized as one of the best penguin exhibits in North America, with six viewing stations to watch the birds swimming and waddling. According to the zoo, "Although it was originally thought that it would take five years for penguin pairs to become established and produce chicks, the first chicks hatched just one year after the exhibit opened. Since then, more than 50 chicks have hatched…an accomplishment for which the zoo is nationally recognized."[10]

In 2007, county lawmakers approved the borrowing of monies for zoo upgrades, including a bigger elephant barn with a spectator pavilion, rehabilitation of the four-acre elephant yard, repairs to the zoo's courtyard and new fencing that created an outdoor exhibit known as Primate Park. Primate Park was designed to teach visitors about the interconnection

An outdoor exhibit known as Primate Park was designed to teach visitors about the interconnection between humans and ecology.

between humans and ecology. Siamangs, ring-tailed lemurs, black-and-white ruffed lemurs and patas monkeys rotated through the space. The exhibit was twelve-sided, thirty feet tall and had mesh covering the sides and top. A viewing cave on one side let visitors get close to the animals. The exhibit contained a pool of water, a small waterfall, grass and plants, tree branches, boulders and six hundred feet of rope for swinging, walking and swooping.

The county authorized $8.5 million worth of renovations and agreed to cover 80 percent of the cost; the $1.6 million balance was to be the responsibility of the Friends of the Rosamond Gifford Zoo. By 2011, the Friends had raised more than half of the money it had promised lawmakers, but it still needed to raise the rest. The Friends launched a community appeal: "Conserving What We Love: The Campaign for Elephants and Primates at the Rosamond Gifford Zoo." The success of the campaign would enable the completion of the zoo's Asian Elephant Preserve, the rehabilitation of outdoor space for primates into Primate Park and the creation of an environmentally friendly courtyard called Gatherings. Furthermore, it would bring home the three elephants who had been sent to the African Lion Safari and Game Farm in Ontario, Canada, during the renovations.

Nancy Rifken and M&T Bank were the major benefactors of the campaign. A new elephant barn was to be named "in recognition of its patrons." "We both always loved animals," Nancy Rifken explained. "I started giving to the elephant house about five years ago, because I thought they really needed a new home. I wasn't happy with the one they were living in." M&T Bank donated $50,000 toward an elephant "biofact" station that would teach visitors about the animals.

A highlight of the campaign was the "Buy a Mile" program focused specifically on the idea of bringing back Targa and her daughter, Mali, the elephants sent to African Lion Safari, and Mali's son, Chuck, who was born at the Canadian zoo. The Safari was 240 miles from Syracuse, and a 480-mile round trip would be required to bring each elephant home, thus racking up a total of 1,440 miles. Donors were asked to "buy" those miles for $10 each, potentially raising $14,400. The fundraiser was a success.

The Jerome C. and Nancy Rifken Family Pachyderm Pavilion, named for its benefactors (*pictured here*), provided a much-needed improvement in elephant housing.

The zoo opened an ocelot exhibit in 2007. "We haven't had ocelots since the 1960s," said Chuck Doyle, zoo director. "We're delighted to welcome ocelots back to the zoo and look forward to aiding in the conservation of these exotic cats." Ocelots are solitary and territorial nocturnal hunters. In low light conditions, their eyesight is six times better than humans'. When Shy and Lisa, a mother-daughter pair, arrived in Syracuse from the Cincinnati Zoo & Botanical Garden, they were kept in a required thirty-day quarantine and then allowed additional time to get used to their new exhibit. Cats were also the featured animals in 2008 when a sand cat exhibit was opened, featuring Kamilah and Chelbi, who were on loan from the Cincinnati Zoo & Botanical Garden. The cats' exhibit was in the Adaptation section in the fennec foxes' former home. The foxes now dwelt across from the ocelots.

Green initiatives undertaken at the zoo in the twenty-first century include switching from single-use plastics (bottles and straws) to aluminum, holding an annual Earth Day Clean Up, posting tips on how to live green at home on social media, using compostable plates and utensils in the café, using products that don't contain palm oil in the café and sourcing local products using vendors who align with the zoo's green initiatives. The zoo is also home to five projects from the Onondaga County's Save the Rain program that collectively keep millions of gallons of water each year from overflowing local storm-water systems and polluting local waterways and Onondaga Lake.

A black-and-white ruffed lemur shows off its bright yellow eyes. Sadly, this species is classified as Critically Endangered by the International Union for Conservation of Nature (IUCN). *Photo by Roxanna Gura.*

The pajama cardinalfish is one of several amazing aquatic species in a shared tank exhibit in the USS Antiquities cave. *Photo by Patrick Lockley.*

Amur tiger mother Zeya patiently puts up with her playful twins, Zuzaan and Coba. This tiger trio represents the largest felid species in the world.

Iniko the patas monkey is very special: Iniko's mother tragically passed away during childbirth, and the zoo's animal care specialists decided to hand-rear her. This was the first documented case of a successful hand-rearing of a patas monkey in an AZA-accredited facility.

A baby yellow-spotted Amazon River turtle represents a species unlike other turtles. This turtle cannot retract its head into its shell and instead tucks its head and neck to one side of its body.

Big Bjorn the Andean bear soaks in the view atop his tree perch. The Andean bear is the most arboreal of all bear species.

Blue crane male Karango and female Karasoki amble through their wetland exhibit. The blue crane is renowned for its brilliant plumage and impressive courtship dance displays.

Bactrian camel brothers Patrick and George. The Bactrian camel is the eighth most endangered large mammal on earth, with fewer than one thousand left in the wild.

Above: The miracle Asian elephant twins stick close to mother Mali. The "Eletwins" are the first recorded Asian elephant twins born outside of their range countries.

Right: Eurasian eagle owl Egon is an ambassador animal, helping the animal care team to inspire appreciation and connect our visitors with nature. *Photo by Kelley Parker.*

The fennec fox boasts massive ears! This small predator uses its keen sense of hearing to locate insects and other prey within a mile.

Two Chilean flamingos form a heart. The zoo proudly hosts a flamboyance—not a flock!—of these stunning birds.

Above: A Humboldt penguin swims in the pool at Penguin Coast. This species takes its name from the Humboldt current flowing along the coast of Chile. *Photo by Michael Villani.*

Left: An endangered neon day gecko displays its impressive ability to stick to tree branches—even in an almost completely vertical position.

Above: Naga the Komodo dragon peaks out through the brush in her climate-controlled outdoor space.

Right: Handsome lion Joshua surveys his territory. The zoo cared for lions until 2018. *Photo by Janette Liddle.*

Left: Muppet the North American porcupine munches on a snack. Muppet is one of the zoo's most beloved animal ambassadors.

Below: Ollie the Giant Pacific octopus shows off his suckers. He has eight arms and nine brains—one in his head and one in each tentacle! This species is incredibly smart and capable of forming bonds with its care specialists.

Above: The Panamanian golden frog is believed to be extinct in the wild, demonstrating the necessity of human care for at-risk species through collaboration with other AZA-accredited facilities. *Photo by Elizabeth Manion.*

Left: A Colobus monkey munches on some of the leaves in Primate Park, while his exceptionally long white tail dangles beneath him. *Photo by Patrick Lockley.*

Regal Amur leopard Rafferty in the Zalie and Bob Linn Amur Leopard Woodland. The Critically Endangered Amur leopard is the rarest felid species on earth.

An endangered eastern massasauga rattlesnake lies coiled in its habitat. This serpent can even be found in Onondaga and Genesee Counties in New York State. *Photo by Grace Brennan.*

Right: A red panda snacks on leaves while standing atop a bamboo walkway in its outdoor habitat.

Below: Sahak the Armenian mouflon. Mouflon are listed as vulnerable due to poaching for their horns. The Rosamond Gifford Zoo is leader of the AZA management plan for Armenian mouflon.

Left: A Steller's sea eagle scans the horizon. This "eagle-eyed" raptor is one of the largest raptor species in the world.

Below: Seahorses anchor themselves in their habitat. Seahorses use their tails so as not to be swept away by currents. *Photo by Bob Gates.*

Above: Siamang ape pair Abe and Fatima sit and spend some quality time together in Primate Park.

Left: A Hoffmann's two-toed sloth hangs from a tree branch. This species spends most of their life in trees, where they can sleep for twenty hours a day!

Above: The majestic snow leopard makes its home in the Himalayan mountains—where conservationists work to save these vulnerable cats from human/animal conflict and habitat loss. *Photo by Daniel Swan.*

Left: Thor the Turkmenian markhor boasts an impressive set of horns. This species' name translates to "snake-eater." The name may derive from their serpentlike horns. *Photo by Scott Reyes.*

Right: Victorian crowned pigeon Saphira sticks close to parents Phil and Lil. They are members of the largest pigeon species in the world, found only in Papa New Guinea.

Below: Turmeric, the male northern tree shrew, is native to the rainforests of Southeast Asia. *Photo by Jennifer Peiffer.*

These projects include porous pavement throughout the zoo's parking lots and the zoo courtyard that prevents rainwater runoff, rain gardens around Primate Park, rain barrels and cisterns to harvest rooftop runoff, wetlands on the Wildlife Trail and around the Waterfowl Pond and a nine-thousand-square-foot Green Roof atop the elephant barn. The zoo also boasts an eco-friendly watering hole at the Helga Beck Asian Elephant Preserve. The fifty-thousand-gallon elephant pool has a biofiltration system that produces zero impact on municipal storm and sewer systems. The Men's and Women's Garden Club of Central New York also maintains a pollinator garden on the Wildlife Trail that is a certified Monarch Waystation. The club tends the garden spring through fall of each year, and it attracts many butterflies and bees while educating the public about the importance of pollinators.

Primate Park, the zoo's outdoor play space for its primate species, opened in 2010. Spring, summer and fall, weather permitting, the zoo's siamang apes, patas monkeys and Colobus monkeys take turns using the complex exhibit for outdoor activities and enrichment.

In another exciting development, Araña, a baby two-toed sloth, was born in 2013. The zoo's first baby sloth in sixteen years was hand-reared by zookeepers. Zoo officials said that Araña was the first Hoffmann's two-toed sloth to be successfully raised by hand in the United States.

The Beck family provided major financial support for the expansion of the Asian elephant exhibit to enable the elephant breeding program to grow.

Henry and Helga Beck always loved animals and frequently brought their three young sons to the Burnet Park Zoo to see them. Henry Beck explained that he had always liked elephants because "they appeared to be smart and caring for others, especially their young." The Becks learned about the zoo's need for funding for the Asian elephant preserve and decided to help support the project. Friends president Janet Agostini was thrilled with the Becks' "immense generosity" in making the largest individual gift in the history of the Friends organization. Agostini was really moved "when Henry asked that the naming gift be listed in Helga's name to honor his wife's contributions and hard work as she stood by him during the growth of their business, Tessy Plastics." "Every donor's concern is that their funds are creating a positive impact," said the Becks. "The whole zoo team proved to us, our family and friends, that the funds were used wisely and effectively. The elephants are enjoying the best possible care and environment."

Ted Fox came to the RGZ in 1991 as a zookeeper. Eight years later, he became collection manager for the bird department, and in 2006, he was named zoo curator. A graduate of Cornell University, Fox holds a bachelor's degree in animal science with a minor in poultry science and a concentration in natural resources. He also taught an honors course, Challenges of Zoo Management, at Syracuse University.

"Fox sits on the Association of Zoos and Aquariums (AZA) passerine Taxon Advisory Group (TAG), collaborating with zoo experts across the country on the conservation needs of this expansive order of birds. Fox was also instrumental in raising the first Andean condor chick to be used for conservation education in Venezuela by the Cleveland Metroparks Zoo in conjunction with the Bioandina Foundation and the Venezuelan National Park Service."[11]

Ted Fox came to the zoo as a keeper in 1991. Fifteen years later, he was named zoo curator, and in 2011, he became the zoo's executive director.

While Fox had loved animals, particularly birds, since he was young, he didn't know much about zoos until 1990, when a friend invited him to visit the Burnet Park Zoo. Ted met bird curator Ken Reininger, who encouraged him to volunteer at the zoo a few days a month. After graduating from Cornell University in 1991 with a bachelor's degree in animal science, a minor in poultry science and a concentration in natural resources, Fox became a part-time zookeeper in the bird department at the zoo. "That's when I started thinking, 'Boy, this field is something really special and it checks so many boxes for me personally,'" he recalled during an interview. It combined many of his interests: conservation, public education and the opportunity to connect with the community.

In 2011, Fox was named zoo director.

When the zoo's new elephant exhibit opened in 2011, it welcomed back two of its female elephants, Targa and her daughter Mali and Mali's three-year-old son, Chuck, who had been sired by Rex in Canada. Targa and Mali had been in Ontario for five years, while four elephants had remained at the zoo: females Siri, Romani and Kirina and Indy, a bull elephant. The returning pachyderms were welcomed with a new barn.

The new structure had viewer windows so the public could see the elephants inside. It measured 11,700 square feet, more than tripling the size of the old 2,500-square-foot barn. Located on the Wildlife Trail, the Rifkin Family Pachyderm Pavilion had a "green" roof, "layered with soil and planted with drought-resistant vegetation to reduce energy use, improve air quality and lessen storm water runoff."[12]

Tiger triplets were the highlight of 2011. The cubs, two male and one female, were born to mother Tatiana and father Toma. "Tatiana is an excellent mother," said Tom LaBarge, the zoo's curator of animals. "With the exception of veterinary health checks, we'll allow her to take care of the cubs without interference." Syracuse.com reported:

> *The litter is Tatiana's second but her first with Toma, to whom she was introduced in December after his arrival from the Buffalo Zoo, officials said. The 11-year-old Tatiana had her first litter, also triplets, in 2004. Of those offspring, brothers Korol and Kunali now live at the Alaska Zoo in Anchorage and their sister, Naka, resides at the Beardsley Zoo in Bridgeport, Conn. Amur tigers are native to eastern Russia and are critically endangered in the wild. Tatiana's grandmother was killed by a poacher and her son, Tatiana's father, was rescued as an orphan cub.*

The zoo started hosting wedding and baby showers and birthday, anniversary and graduation parties in 2004, offering a "wildly unique" setting for groups from fifty to three thousand. Their advertisements were enticing:

> *Minutes from downtown Syracuse and surrounded by tree-covered hills, the Rosamond Gifford Zoo is not only filled with beauty, but also inhabited by*

Above: Triplet Amur tiger cubs were born in 2011 and were exhibited to the public via a live video feed from the cubbing den.

Left: The birth of critically endangered snow leopard cubs in 2012 made the RGZ one of only four zoos in the country to have had new births of endangered animals.

Opposite: The zoo offers a unique setting to celebrate very special nuptials, with proceeds benefiting the animals and supporting the zoo's mission.

some of the most intriguing and delightful animals you could ever want to meet. They were probably among your first friends, so why not include them at your most important events?

Animal-specific venues included the main overlook of the Helga Beck Asian Elephant Preserve, which "offers an awe-inspiring view of magnificent Asian elephants interacting in the preserve just below," or the patio at the Humboldt Penguin Coast, which "offers a unique spot to say your vows in front of the best-dressed birds at the zoo." The event could be made even more memorable, the zoo suggested, "by inviting one of our charming animal 'ambassadors' to meet and mingle with your guests."

In 2012, after a fourteen-year hiatus, snow leopard cubs were again born at the Rosamond Gifford Zoo. Ruka was the first snow leopard cub born at the zoo, in 1997, and Tian and Nema were born the following year. The new cubs were the first litter from Zena and Senge, two of the four snow leopards who live at the RGZ. "They are kittens now, but we won't be able to handle them for much longer," said director Fox as the cubs were shown to the public for the first time. "A couple more days is all we have," he said. "They are like any other wild animal." Snow leopards are critically endangered in the wild, and according to Fox, at the time, there were only 137 snow leopards in 63 zoos in the United States; only 4 other zoos in the country had cubs born that year. The cubs were named Bajen, which means "long-awaited," and Ramil.

In 2012, the Rosamond Gifford Zoo hosted the Thirty-Ninth Annual American Association of Zookeepers National Conference. The conference schedule offered eighteen papers on topics such as training, conservation, innovations and enrichment. Twenty-two workshops highlighted husbandry and management programs, zoonotic disease and quarantine protocols, team-building strategies and techniques for creating a successful AZA chapter. Zoo Day at the RGZ featured training demonstrations and behind-the-scenes tours. The end of the conference featured a Bowling for Rhinos Pep Rally that brought together leaders in conservation, including representatives from the Lewa Wildlife Conservancy, the International Rhino Foundation and Action for Cheetahs in Kenya.

In 2013, a strategic master plan for the zoo was created. The Friends of the Zoo engaged a company called Zoo Advisors to assist in the development of a comprehensive integrated strategic plan to guide the zoo's future. Zoo Advisors partnered with GLMV Architects to conduct the four-phase project, with a planning team composed of representatives from the county, Friends, zoo leadership and staff and members of the zoo's board. The team conducted a thorough review of the site and facilities, held numerous design workshops and concluded with concepts for exciting new exhibits and a flow intended to make the best use of the zoo's site and "viewscapes." The ultimate goal was to leverage past success, generate excitement around a new and compelling mission and vision, attract broader audiences from throughout Central New York, engage them in new and innovative ways and guarantee them the "best day ever."

The twenty-year strategic plan included a combination of short- and long-term projects and a business plan. Zoo Advisors and GLMV Architects worked for more than a year with a planning team that included representatives from Onondaga County, Friends and Zoo leadership and board and staff members. A "Blue Ribbon Panel" brought thought leaders from around the country to the Zoo for a "blue sky session." Officials hoped that the new plan would keep the zoo on solid footing, improve the guest experience and make the zoo a regional destination.

The plan assessed strengths, weaknesses, opportunities and threats. Strengths included the partnership between the county and the Friends organization; the sharing of governance; the zoo's board, staff and volunteers; its reputation, consistent value and location in Central New York; and its financial soundness. Weaknesses listed were the zoo's location and lack of accessibility, its staffing capacity, difficult navigation of the zoo for visitors, entrance bottlenecks, logistics barriers and the lack of visitor awareness of the conservation program.

The report listed several opportunities: bringing in more off-season guests, available expansion space, contact barn renovation, area development (downtown, Destiny, CVB), Syracuse University, SUNY, the emerging Green Hub, STEM and senior and young adult markets. Simultaneously, it recognized potential threats: the perception of the zoo's neighborhood as dangerous, the competition with Destiny and its proposed aquarium, the availability of animals in the future, nearby non-AZA facilities offering contact experiences and the negative image of the city of Syracuse itself.

A new mission statement was developed: "The Rosamond Gifford Zoo is dedicated to connecting people to the natural world by providing engaging guest experiences, exceptional animal care and unparalleled conservation education." The mission statement was accompanied by a vision statement: "Rosamond Gifford Zoo provides its guests the 'Best Day Ever,' ensuring experiences that excite, memories that endure, and knowledge that inspires worldwide conservation."

The site plan, it was emphasized, was a twenty-year vision for the zoo, encompassing "big picture" ideas. Exhibit specifics were subject to change, including the exact animal species to be displayed. The team evaluated and planned for a couple of areas to minimally grow outside the zoo's perimeter to help with parking and guest flow. Phasing and sequencing of the plan could change based on a variety of factors, including funding, donor interest, collection planning, county priorities or AZA requirements.

Six core values were identified:

1. *Animal Welfare: We value animal well-being above all else and make all decisions with animal health and safety as a top priority.*
2. *Fun: We value fun and believe that fun is a gateway for learning and that it is essential for a healthy workforce.*

3. *Sustainability: To be a truly sustainable organization and to educate and encourage our guests to conserve, we must first model sustainability ourselves by saving resources and making wise choices in our business and operating practices.*
4. *Exceeding Standards: We go above and beyond in all we do, surpassing expectations of our guests and exceeding standards of our profession.*
5. *Integrity, Honesty and Trust: We value integrity, honesty and trust, and model these values for our guests, our business associates, the community and each other.*
6. *Diversity and Inclusion: We value diversity and believe that inclusion of all viewpoints and perspectives will enrich our workplace and help our Zoo grow and prosper.*

The master plan was comprehensive, detailed and forward-looking, providing a clear roadmap for decades. Its authors concluded,

> *The time is right to take that development at the Zoo to the next level. A compelling vision has been laid out to chart the next round of Zoo improvements with the first projects already identified. With the "can do" attitude found both at the Zoo and the support of the Zoo's many stakeholders, the innovative exhibits ideas can move from paper to reality. This is a time of great excitement for the Zoo, yet one that will require the leadership and support of many: the Zoo staff and Zoo leadership, the Board, the County and other community stakeholders. It is only with everyone working together that this vision can be realized.*

The master plan included the creation of a new exhibit featuring a giant Pacific octopus, the largest and longest living of the octopi. Found in the North Pacific Rim, they can grow up to twelve feet long and weigh up to fifty pounds, though they do not usually get that big. A Florida company designed the 1,600-gallon exhibit tank. As Syracuse.com described it,

> *The exhibit was a big hit with the 350,000 people a year who visited the zoo. For one thing, octopuses look really odd. They have big heads and four sets of arms, each lined with two rows of suckers that help them detect and capture prey on the bottom of the sea. To get blood to all those parts, they have three hearts. Their bodies have no bones. So even big ones can squeeze through the tiniest of openings in rocks and caves. They have excellent eyesight, an outstanding sense of touch and are superb escape artists.*

The Iorio family (*pictured*) sponsored a 1,600-gallon saltwater aquarium, which became home to Ophelia, a giant Pacific octopus.

Ophelia the octopus debuted in the summer of 2013 in a 1,600-gallon saltwater tank at the entrance to the zoo's USS Antiquities area. Several species of fish, sea stars and anemones also resided in the tank. Thousands of visitors were intrigued and mesmerized by the exhibit, the first ever octopus exhibit in upstate New York, which was funded by Friends of the Zoo through a private donation from Rick and Laura Iorio of Manlius. The Iorios have been members of the zoo since their children were young and say that visiting the zoo and providing support are family traditions. "Our son, Jason, used to love zoo camp," recalled Laura. "I can still remember how much he loved studying the sloths and telling me what he helped make them for lunch." "The zoo is doing such a fantastic job in caring for the animals. Just look at all the healthy animal births, as well as the zoo's AZA accreditation and numerous recognitions," said Rick. "The zoo does a great job of not only educating the children who come to visit but of making it a fun, positive experience for everyone."

Ophelia was caught in the North Pacific and brought to the zoo weighing about 7 pounds. Zoo director Fox said she would probably grow to about 40 pounds in the tank, although in the wild, the giant Pacific octopus can reach 110 pounds and sixteen feet in length—each of its eight arms may measure over six feet. Female octopi can lay one hundred thousand eggs at a time, but there were no plans to breed Ophelia because the giant Pacific octopus is notoriously difficult to breed in captivity. Ophelia's lifespan was expected to be four to five years. The tank, designed by Living Color, was featured on an episode of National Geographic Wild's *Fish Tank Kings* in April 2014.

The zoo celebrated its one hundredth anniversary in 2014. When the Burnet Park Zoo opened in 1914, it consisted of a barn, four acres of land and fewer than a dozen animals, mostly local species. One hundred years later, the zoo covered forty-three acres and had seven hundred animals, consisting of 240 species. To celebrate that achievement, on July 14, 2014, the zoo held an anniversary celebration. Visitors signed a giant birthday card, and county executive Joanne Mahoney read a proclamation and said, "One hundred years really means a lot to us because it demonstrates the support of this community. We're constantly adding new exhibits, another thing that some zoos just haven't been able to do, as the money just isn't coming in from the community to create those experiences. We're adding something every year, sometimes multiple things in a year." Admission to the zoo on the anniversary was only a dollar, and there were daily drawings during the summer for behind-the-scenes experiences at the Asian elephant, Amur tiger and Humboldt penguin exhibits. Mahoney also announced that all veterans, active-duty military members and their immediate families would receive free admission to the zoo.

Batu, a male Asian elephant calf, was born on May 12, 2015, making the Rosamond Gifford Zoo the only zoo in North America to have three generations in one herd. That year, the zoo was given the new Quarter Century Award, presented by the Association of Zoos & Aquariums. The award recognizes zoos and aquariums that have been AZA accredited continuously for twenty-five years or more. "Accreditation is our guests' guarantee that we abide by the highest standards in animal care, conservation education and visitor experience—a responsibility we take very seriously," said the county executive. "We are pleased to be recognized by the AZA for our dedication to excellence and to our mission. Congratulations to the zoo and its staff for this achievement."

According to Janet Agostini, "The zoo provides an excellent opportunity to ignite someone's interest and concern for the natural world"—seeking to ignite interest and concern for the natural world both from education and from play.[13] Thus, in 2015, the Friends of the Zoo funded a $113,000 renovation of the zoo's gift shop. The Curious Cub Gift Shop was reshaped as an elongated D. The right wall, made entirely of windows, showcases a tremendous variety of plush animals and highlights winning images from the annual Winter at the Zoo Photo Contest. The shop stocks items

Left: Batu's birth in 2015 made the Rosamond Gifford Zoo the only zoo in North America to have three generations of elephants in one herd.

Below: The Curious Cub Gift Shop offers a variety of items for animal lovers, and purchases benefit animal care and conservation at the zoo.

representing species exhibited at the zoo—including specialty items like Elephant Poo Paper, stationery products produced from elephant dung.

> *The zoo's mission is to provide an engaging guest experience, and our vision is to provide the "best day ever" for everyone who walks through our doors. The Curious Cub Gift Shop serves as the last touch point for many of our guests, and we take great pride in providing a great experience with*

> *knowledgeable staff to set the stage for their return visit. Engaging each guest, making them feel welcome and helping them have fun while exploring the store are all important components to creating an atmosphere people want to return to and support.*[14]

Red panda cubs Pumori and Rohan were born at the zoo in 2015. Their mother, Tabei, was a two-year-old first-time mother; their father, Ketu, was a four-year-old first-time father who came to Syracuse from New Zealand to diversify the genetic pool of the North American red panda population. Red pandas are endangered, with fewer than ten thousand in the Eastern Himalayas and Southwestern China. The loss of nesting trees and bamboo due to deforestation has caused a decline in their numbers. Red pandas are thought to be distantly related to both giant pandas and raccoons, but they have no close evolutionary relatives; they are a unique species. "It is always exciting to have new babies at our zoo. These red panda cubs are important to the North American population and a testament to the hard work of our zoo staff. I commend the dedicated keepers and veterinarians for their continued success," said county executive Mahoney.

In 2016, the zoo completed several large projects to enhance and improve facilities for both animals and visitors. The most important project was the completion of the elephant pool at the Asian Elephant Preserve for the zoo's seven pachyderms. The pool mimics a natural watering hole and has observation decks for visitors and a life support system for maintaining water filtration with green infrastructure. It allows the elephants to engage in wallowing, wading, splashing, cooling and bathing and assists with skin care and protection from the sun and biting insects.

In 2016, the zoo opened a new nature play space at Explorer's Outpost, offering a nature play area for children with a dig/build area, a faux log tunnel, a willow hut, a marimba, a shark-shaped kitchen and more. Kids were invited to explore the dig zone, crawl through the giant log tunnel, play a tune on the marimba, grab some shade in the willow hut and splash water on their faces from the bucket-filling station. Friends of the Zoo worked on the play space with Rusty Keeler, author of *Natural Playscapes* and designer of children's environments, who has led a movement to reconnect kids to nature through play. Initial funding for the space was provided by M&T Bank as part of a capital campaign gift that supported construction of Explorer's Outpost, with additional support from Friends of the Zoo.

A sculpture of the Chittenango ovate amber snail by local artist Kate Woodle was funded by a grant from the U.S. Fish and Wildlife Service to

The elephant pool at the Asian Elephant Preserve mimics a natural watering hole and has observation decks for visitors. *Photo by Erin Fingar.*

Explorer's Outpost is a child-sized building where kids can explore the differences between different kinds of animals.

Left: The Chittenango ovate amber snail, an endangered snail found only in nearby Chittenango Falls State Park, is part of the zoo's breeding and release program.

Right: In 2017, Siri, the beloved and oldest elephant resident of the zoo, turned fifty, and the zoo celebrated with a series of elephant-themed events called Pachyderm Parties.

spread awareness of the tiny, endangered snail found in only one place in the world. The Chittenango ovate amber snail (COAS) is a rare and tiny inhabitant of Central New York. In fact, the only place in the world where this species exists is Chittenango Falls State Park. The thumbnail-sized land snail lives in the mist zone of the 167-foot waterfall for which the park is named and feeds on decaying leaves. Only about three hundred snails exist in the wild.

Since they exist in such small numbers and live in only one place, Chittenango ovate amber snails are extremely vulnerable to extinction. Invasive species of plants are encroaching on their habitat and may supplant the native vegetation the COAS needs to survive. Climate change and human encroachment on their habitat are also threats. The Rosamond Gifford Zoo is part of a coalition of conservation organizations that have teamed up to save the snail. Called the Snailblazers, the group includes researchers from the State University of New York College of Environmental Science and Forestry, who have pioneered hand-rearing the snails; the U.S. Fish and Wildlife Service; the NYS Department of Environmental Conservation; the NYS Office of Parks, Recreation and Historic Preservation; and the Seneca Park Zoo in Rochester. The recovery plan for the snails included establishing captive colonies to provide genetic backup for the wild population, were it to diminish, as well as increasing the wild population through releases of captive-raised snails. Research supported by the Great Lakes Restoration Initiative provided data about the snails' dietary preferences and husbandry that has enabled the Snailblazers to maintain populations in terrariums kept in temperature-controlled incubators housing some four hundred snails in colonies at the zoo and at SUNY-ESF.

After a $300,000 renovation, the zoo's banquet facility and patio, completed in 2016, quickly became a popular wedding spot, hosting a record-breaking thirty-four nuptials in its first year. The upgrade made possible such popular events as birthday parties, Mother's Day Brunch and Breakfast with Santa.

The 2017 Summer of Siri was a celebration of the nine-thousand-pound elephant's fortieth birthday. Pachyderm Parties with games, face-painting and crafts delighted young guests, and the zoo offered guided walks of the Asian Elephant Preserve, where visitors could watch the whole elephant herd, including Siri, Doc, Romani, Kirina, Targa, Mali and Batu, enjoy watermelon, their favorite summer snack. A weeklong series of elephant-themed activities culminated with the Asian Elephant Extravaganza, a popular event celebrating both Asian elephants and the traditional cultures of their native countries. Siri was painted with an Asian headdress and enjoyed a special birthday cake. The zoo held a Pennies for Pachyderms fundraiser to aid Siri's counterparts in the wild, raising over $7,000.

In 2017, the zoo's herd of seven elephants was joined by an eight-month-old Rottweiler puppy whose job was, as the local paper reported, to "provide enrichment for the elephants and serve as an ambassador for the zoo." Named Moakler, the puppy was introduced to the elephants when he was eight weeks old. "[Zoo director Ted] Fox said pachyderms are intelligent animals that, in the wild, would come in contact with lots of other animals." "We thought it'd be really interesting to have them be stimulated," Fox said. "The trainers and our visitors to the barn interact with the dog and are enriched and have something to watch." "Fox said for the animals, it's about training them to be as comfortable as possible around people. Staff see how animals learn obedience." While Moakler was the zoo's "first official elephant barn dog," other dogs have worked in the contact barn, Fox said.

Another new addition to the zoo in 2017 was its first American bison baby, born on the Fourth of July. The zoo celebrated the newly designated national mammal with a Baby Bison Bash on July 5, where guests voted to name her Abigail, after First Lady Abigail Adams. The party featured free American flags, face-painting and a bison-themed raffle basket. There was also a special bison keeper chat in the afternoon, where visitors could learn fun facts about our national animal, including that bison are native to North America, buffalo live in Africa and South Asia and the two are distinct species of the same family.

In 2018, two Bactrian camels joined the zoo family. Half brothers Patrick and George came from the Milwaukee County Zoo at the age of six months, weighing around six hundred pounds each. Three years later, they weighed two thousand pounds. Bactrian camels have two humps and are acclimated to the cold temperatures of Mongolia's Gobi Desert, in contrast to their one-humped relative, the dromedary, native to the hot Sahara Desert. They are critically endangered in the wild due to hunting and competition with other livestock for food. Fewer than one thousand individuals remain in their native habitat in the Gobi Desert in northern China and Mongolia, making them the eighth most endangered large mammal in the world. The Wild Camel Protection Foundation is working to save them from extinction through a breeding program at the Hunter Hall Wild Camel Breeding Centre in Mongolia.

Red panda cubs Loofah and Doofah were born at the zoo in June 2018. The brothers were born to the zoo's resident breeding pair of red pandas, mother Tabei and father Ketu, and were hand-raised by keepers "after Tabei demonstrated some difficulty in caring for them on her own."[15] The keepers bottle-fed them every four hours and monitored their intake and weight. According to the website ZooBorns, "Red pandas are an endangered species, with fewer than 10,000 estimated remaining in the wild in the Himalayan Mountains. They are called pandas because, like the Giant Pandas of China, they eat primarily bamboo. The word 'panda' comes from a Nepali word meaning 'bamboo eater.'"[16] The RGZ was involved in increasing the red panda population through the Species Survival Plan (SSP) for red pandas overseen by its accrediting organization, the Association of Zoos & Aquariums (AZA). Zoo director Fox said that the twins were Tabei's third set of cubs since 2015. Her first cubs, males Rohan and Pumori, went on to start their own families at the Central Park Zoo and the Erie Zoo. Ravi and Amaya, a male and female born in 2016, are now at the Detroit Zoo and the Sacramento Zoo, respectively. The red panda SSP works to pair unrelated animals from a diverse gene pool in the interest of producing healthy offspring for survival of the species.

Dinosaurs invaded the zoo in the summer of 2018. A dozen colossal animatronic creatures were on display on the Wildlife Trail and in other outdoor areas, including a twenty-three-foot-long Baryonyx, 40-foot-long Tyrannosaurus rex and two-story-tall Brachiosaurus. "We are so excited to be able to present 'Dinosaur Invasion: 101 Days of Dinosaurs,'" said Friends director Janet Agostini, quoted on Syracuse.com. "We have seen these dinosaurs at other zoos, and there's something so compelling about

A dozen colossal animatronic dinosaurs invaded the zoo in the summer of 2018.

encountering them outside in a natural setting." "Zoo director Fox said he was thrilled when the Friends offered to bring dinosaurs to the zoo… [because] it presents opportunities for the zoo to talk about the animals in its collection that share characteristics with dinosaurs, explore the link between dinosaurs and birds, and illuminate the zoo's role in saving today's endangered species from extinction."

Asian elephants are the species the Rosamond Gifford Zoo is most proud of helping to save as part of its AZA wildlife conservation mission. Of several thousand zoos and aquariums in North America, only 232 have passed the rigorous inspections required for AZA accreditation—and of those, only 30 have Asian elephants and only 11 have breeding programs. Asian elephants are critically endangered in Asia and India due to habitat destruction, poaching and hunting, and there are only thirty thousand remaining in the wild.

In January 2019, the zoo made history when an elephant calf arrived early one morning—the second calf born to Mali and Doc, both twenty-one years old. The baby weighed 268 pounds and measured about three feet tall at birth. "Asian elephants are critically endangered in the wild, so it's a huge accomplishment to be able to breed them in human care," noted county executive McMahon. "I congratulate the zoo and its dedicated animal care

Asian elephants are the species the Rosamond Gifford Zoo is most proud of helping to save.

staff, as well as the Cornell University College of Veterinary Medicine team that assisted them in preparing for this birth."

As reported in the *Citizen*, Mali's pregnancy lasted an estimated 660 days. After determining that her pregnancy was progressing well, the elephant care and veterinary team began preparing a year in advance. They conducted birthing drills in the elephant husbandry barn, using a life-size inflatable elephant to represent Mali and a giant boat buoy to represent the baby. When Mali started showing signs of active labor, the team was ready when the male calf arrived less than half an hour later. "Mother and baby are both doing fine," reported Fox, adding, "We will monitor them closely while giving Mali and her newborn time to bond."[17]

As reported by Onondaga County Parks,

> *The zoo posted photo and video updates on its social media platforms so the public could see the baby's progress leading up to a springtime introduction to the public. The zoo is undergoing a construction project to expand the Asian Elephant Preserve from 4.5 acres to 6 acres and improve viewing access to elephants and other species on the Wildlife Trail. The construction is expected to be completed by Memorial Day weekend.*
>
> *"We are looking forward to a great summer filled with opportunities to watch this little baby grow,"* [zoo director] *Fox said....*
>
> *The new addition brings the zoo's elephant herd to 8 animals, including a three-generation family group that includes Mali and Doc's first calf, Batu, a male who turns 4 in May, and Mali's mother, Targa, 35. The Preserve also is home to the calves' three unrelated "aunties"—matriarch Siri, who turns 52 this year, as well as Romani, 41, and her daughter Kirina, 23.*[18]

The zoo's new elephant walkway and overlook opened in 2019. The $2.2 million walkway allowed visitors to get the first sighting of five-month-old, seven-hundred-pound baby elephant Ajay. Zoo director Fox said that the new exhibit was one of the best in the country in terms of size and complexity. "We put elements in that don't exist anywhere else, like the walk-through tunnel, the topography and the grass seed mix. There are very few places in the country where you can see elephants grazing on natural grass, and that's exactly what they get to do here." Fox added that "there are only thirty zoos in the country that have Asian elephants. Only eleven have breeding families, and we're one of those eleven." When the reservations were complete, the preserve would grow from four and a half to seven acres for the herd of eight elephants.

Seven giant animatronic insects descended on the Wildlife Trail of the zoo in the summer of 2019. "Our Big Bugs are here to serve the same purpose as our dinosaurs did last summer—to spark interest in science and natural history, especially among our younger guests," said Janet Agostini, president

Giant animatronic insects, like the dinosaurs that preceded them, were on display on the Wildlife Trail in 2019 for the edification and delight of zoo visitors, young and old.

of the Friends of the Rosamond Gifford Zoo, which sponsored the exhibit. The giant bugs were a seven-spotted ladybird, a Madagascan sunset moth, a blue-eyed darner, a bombardier beetle, a meadow grasshopper, a red-tailed bumblebee and an orb-web spider.

Perhaps even more thrilling was the birth, in early May, of a baby bison. American bison once numbered in the millions but came close to extinction by 1900, when only about one thousand remained. Conservation efforts have restored the bison population to about five hundred thousand in zoos, preserves and protected parklands. "With our second bison birth, we are doing our part to contribute to the health of this species," said Zoo director Fox. "It is a great experience to participate in the conservation of this iconic animal."

Another new addition to the zoo was an eighteen-horse and one-chariot carousel, installed by the Friends of the Zoo in the green space across from the Outdoor Birds exhibit. The colorful musical attraction cost $65,000 and lent a festive, nostalgic touch to a parklike area of the zoo. Rides were $3 each, and proceeds benefited the zoo's animals. The outdoor ride was open daily, weather permitting, and was a bit hit with young and old alike from the start.

In 2020, in celebration of the zoo's fiftieth anniversary, the Friends of the Zoo announced a campaign to raise "$50K for 50 Years." "Just as the zoo has changed a lot in 50 years, so has the Friends. We have gone from a small group of citizens that formed to save a failing zoo to a key partner of the zoo that supports its AZA missions of top-notch animal care, nature education and saving species," the Friends appeal stated, noting that no gift was too small. The funds raised were destined to help create a state-of-the-art animal health center not only to provide medical care for the animals but also to serve as a teaching facility for future zoo veterinarians.

The Friends described the need:

> *When the Rosamond Gifford zoo's animal health clinic was built in the 1980s, it was ahead of its time period. Besides providing an exam room, X-ray station, operating room and quarantine rooms for zoo animals, our clinic was among the first to have public windows offering zoo visitors a live view of animal health care and action. That was 40+ years ago. Since then the number of animals in our care has grown to include many more threatened and endangered species. We've seen huge advances in medical technology and zoological medicine. Our participation in global Wildlife Conservation projects and research has intensified and we now must consider how the health of animals raised in our care may impact how we can introduce them into the wild.*
>
> *Our little clinic has become woefully inadequate to the missions at hand. In our last two accreditation cycles in 2013 and 2018, the Association of Zoos & Aquariums (AZA) noted that our clinic is far too small for a zoo of our size, cannot provide hospitalization for most of the larger species in our care and lacks proper quarantine space to meet best safety practices. Our AZA accreditation is extremely important—it puts us on the map of only 240 zoos and aquariums considered the gold standard for animal care, Wildlife Conservation and education, period—but that's just one reason we are about to build a new animal health center. Our zoo is devoted to providing the very best health care available to every animal in our care, from one tiny poison dart frog to a 10,000-pound elephant—and to maintaining our growing reputation as a leader in animal care and welfare in collaboration with one of the world's top veterinary schools. A new state-of-the-art health center will ensure we meet those standards for many years to come.*

In 2020, Friends of the Rosamond Gifford Zoo also unveiled the new habitat for the zoo's Amur leopards, the Zalie and Bob Linn Amur Leopard Woodland. As the zoo announced at the time, "The new exhibit, located in the zoo's former African lion exhibit, provides an enriching outdoor space for the world's rarest big cat species."[19] Friends president Agostini said, "We are really thrilled that the new woodland habitat will be named for a couple whose love for animals and for their community goes above and beyond." The new project renovated the former African lion exhibit at the zoo for Amur leopards, a critically endangered species of big cat native to the forests of Siberia. The exhibit was to be shared on a rotating basis with another

The Zalie and Bob Linn Amur Leopard Woodland provides an enriching outdoor space for the world's rarest big cat species. The Linns are pictured here.

endangered species, ruffed lemurs, according to Zoo director Fox. The zoo began planning for a new exhibit when it joined the Association of Zoos & Aquariums' Species Survival Plan for Amur leopards in 2018.

According to the Linns, they signed on to support the project as a gift to the zoo and their community. Bob Linn said, "I think our community should be very proud of what the county, the Friends of the Zoo and its many volunteers have achieved."

According to the zoo,

> *The exhibit is now home to the zoo's breeding pair of Amur leopards, Tria and Rafferty, and their offspring. The pair produced twin cubs within a year of being introduced at the zoo. Their cubs, Milo, a male, and Mina, a female, were born in June 2019 and will be moving to other Association of Zoos & Aquariums institutions as part of the Species Survival Plan to save Amur leopards from extinction.*
>
> *Amur leopards are critically endangered, with fewer than 60 estimated to remain in the wild. Accredited zoos care for an assurance population of about 200 through the Species Survival Plan. Amur leopards also are a cold weather species, so they are perfectly acclimated to Central New York winters and will be able to use the outdoor habitat year-round, Fox said.*[20]

The COVID-19 pandemic had a major impact on the zoo. Maria Simmons, the zoo's director of marketing and communications, recalls,

The Friends of the Zoo entered 2020 with high hopes and exciting plans for the zoo. However, everything changed on Tuesday, March 16. As was true for many organizations, companies and workplaces, the zoo had to close abruptly when COVID-19 struck our region. When Onondaga County closed the zoo to all but essential county staff that day, the Friends team pivoted to work remotely from their homes for as long as would be needed. The zoo increased communication efforts to keep members, volunteers, the community, and tens of thousands of social media followers abreast of what was happening and how the animals were faring during the closure. Special programs were developed and designed to keep constituents connected and engaged during the shutdown. A Learn at Home web page was created to assist families who suddenly found themselves home with children whose schools were closed. The Learn at Home page provided keeper chats as well as activity and craft videos contributed by our educators. The Friends education team worked to reformat programs and hone their video presentation skills to reach young learners online both at the local level and beyond. From July through December, the team provided 63 Zoo to You virtual programs seen by 2,043 participants in three states and Canada.

Then, on May 23, 2020, the Rosamond Gifford Zoo became the first zoo in New York State to reopen, thanks to an intensive plan submitted by the zoo and approved by the county executive and the county health department. New safety protocols went into effect on Memorial Day weekend. The zoo constructed plexiglass safety barriers at the membership desk and ticket booths; configured a one-way touring route through the outdoor zoo; implemented masking and social distancing requirements enforced with signage, ground and floor directionals and safety tape; and enforced an attendance cap of 15 percent capacity for most of 2020. When the zoo reopened, social media updates were crucial in informing potential zoo-goers which exhibits were open, how to make reservations and what the COVID protocols were. As part of the reopening, regional billboards carrying the message "We miss you, too"—with a photo of a red panda—conveyed a heartfelt message to the community and were warmly received.

Tragically, in 2020, death came to the new zoo just as it did to the old zoo. In January 2020, the zoo announced the sudden death of its youngest elephant calf, Ajay, from EEHV, a lethal strain of herpes that targets Asian elephants.

Ajay, who would have turned 2 this coming January 15, was the beloved baby of the zoo's elephant herd and the second calf born at the zoo to his mother, Mali, and father, Doc....

EEHV, or elephant endotheliotropic herpes virus, is the biggest killer of young Asian elephants and can cause death within 24 hours in those under age 8. EEHV was discovered by the Smithsonian National Zoo Conservation Biology Institute in 1995 after the National Zoo lost a 16-month-old elephant calf to the disease. The Smithsonian established the National Elephant Herpesvirus Laboratory as a result, and the Rosamond Gifford Zoo is a member institution that sends blood samples from its elephants to the lab twice weekly to monitor for the disease....

Last Thursday, a blood sample from 5-year-old Batu tested positive for EEHV, and the zoo began aggressively treating him although he had no symptoms. Batu remained asymptomatic over the weekend even as subsequent blood tests showed that the levels of virus in his blood were growing exponentially by the day.

While treating Batu, the zoo and its Cornell University veterinary team were closely monitoring his little brother, Ajay. Ajay showed no symptoms until the end, Fox said. On Monday night, he was playful and happy as usual. By Tuesday morning, there was some swelling around his eyes and his tongue was slightly darkened, signs that EEHV is active.

Although Ajay wanted his morning bath and ate some of his breakfast, he seemed tired, Fox said. "Within two hours, despite the tireless efforts of

Batu and Ajay were inseparable siblings whose deaths from elephant endotheliotropic herpes virus within days of each other was a devastating blow to the zoo. *Photo by Erin Fingar.*

> *the zoo and Cornell veterinary staff, he was gone. That is how awful this disease is, especially with young elephants. When it hits, it hits hard."*
>
> *Fox said the entire herd was given the opportunity to spend time with Ajay after his death as elephant families do in the wild....*
>
> *Fox said Ajay has been taken to Cornell University College of Veterinary Medicine and his death will contribute to study of EEHV, which also affects Asian elephants in the wild. He said the loss of Ajay will make a difficult year even harder for the zoo and the many people who love and support its elephants.*

As described on the zoo's Facebook page, "When Ajay was born on January 15, 2019, Batu took an immediate interest and the pair became almost inseparable. The sight of the two brothers playing, wrestling or splashing in the pool together was an adorable highlight for many zoo visitors." So it was even more tragic when Batu succumbed to the EEHV virus three days after Ajay. "It's just an absolute devastating day for the community again, and for our parks team and our zoo team," said Onondaga County executive Ryan McMahon, announcing Batu's death.

According to the zoo's 2020 annual report,

> *On September 10, Onondaga County broke ground on the long-awaited Animal Health Center on a hilltop site adjacent to the zoo parking lot. The county allocated $7.5 million for the health center design, engineering, site work, construction, infrastructure and landscaping, while the Friends committed to contributing $1.1 million to outfit the center with state-of-the-art diagnostic and treatment equipment.*
>
> *The Friends development, marketing and communications departments worked together to create a detailed case statement and collateral materials for a capital campaign to launch in January 2021 with the goal of securing naming gifts, sponsorships and donations for the project.*
>
> *When completed, the health center will be the largest zoological medical center outside of the Bronx Zoo. The 20,000-square-foot health center and quarantine facility will provide top-tier animal care and further the zoo's 22-year teaching partnership with Cornell University College of Veterinary Medicine, as well as enhancing the zoo's visitor education mission.*[21]

On April 12, 2021, Onondaga County executive Ryan McMahon officially opened the new Animal Health Center at the Rosamond Gifford Zoo with a ribbon cutting. The opening came after a decade of planning and a year of construction. The new freestanding zoological medical center replaced a small in-house veterinary clinic. McMahon called the new health center "a historic accomplishment for our internationally accredited zoo and for Onondaga County." "It also offers opportunities for zoo visitors to observe animal health care in action. Expansive windows offer views into treatment rooms, surgical suite, nursery/ICU, animal nutrition kitchen and research lab so visitors can experience the work that goes on behind the scenes for animal care and welfare at the zoo."[22]

"The 'One Health | One Mission' campaign has exceeded its $1.1 million goal due to the support of community members, organizations and foundations, said Friends of the Zoo Executive Director Carrie Large." "Large and Fox said every aspect of the health center is meant to serve the 'One Health' philosophy that humans, animals and the environment are all connected and the health of one affects the others."[23]

"We named our capital campaign 'One Health | One Mission' because the health center will enhance our mission to connect people to animals, nature and our planet," Large said.[24]

"The new health center also meets and exceeds the standards for accreditation by the Association of Zoos & Aquariums (AZA), whose 240 member institutions must demonstrate a commitment to animal care, guest education, wildlife conservation and preserving threatened and endangered species."[25]

"More space and a dedicated laboratory will expand the zoo's ability to treat more and larger animals on-site and contribute to research advancing animal science, Fox added. A separate but adjacent animal quarantine facility will meet best safety practices for bringing new animals into the zoo, with four specialized areas to meet the needs of carnivores, primates, aquatic animals and birds." [26]

"As an accredited member of the Association of Zoos & Aquariums (AZA), the Rosamond Gifford Zoo is dedicated to providing a strong educational component to inspire visitors to care about animal care and conservation," said director Fox. Thus the health center also housed a Junior Veterinary Clinic designed to provide an "imaginative play space for learners of all ages and abilities to engage in pretend animal care using toy medical tools and plush animals."[27]

"The new 'clinic' is a large room off the new Animal Health Center furnished with play stations modeled on a typical veterinary clinic, including waiting room, check-in, exam, diagnosis, surgery and diet.

"Large said having the Junior Veterinary Clinic will encourage kids to put what they see and learn into play. Children can bring their own stuffed animals or choose a toy 'patient' there and use the stations and toy equipment to role-play as customers bringing their pets to the vet; veterinarians or vet techs performing exams, treatment or surgery on toy animals; receptionists checking patients in or out; preparing diets, 'whatever animal care roles they can imagine,' Large said.

"The zoo's education team is also working to devise Summer Zoo Camp sessions that incorporate the Junior Veterinary Clinic and allow young people to meet and learn from Cornell veterinary and zoo animal care staff."[28]

Carrie Large was named executive director of the Friends of the Rosamond Gifford Zoo in 2020, following a national search. "Friends of the Zoo Board Chair Heidi Holtz said the board felt Large was a good fit to lead the Friends organization as it seeks to help the zoo recalibrate after the COVID-19 pandemic, sustain its highly respected reputation in animal conservation and education, and pursue an ambitious strategic and master plan that includes support for a new Animal Health Center now under construction."[29]

"We look forward to working with Carrie, who we believe will bring a powerful mix of professional expertise and personal dynamism to the Friends organization," Holtz said.

Carrie Large was named executive director of the zoo in 2020, just in time to lead an ambitious new strategic master plan.

"My family loves our zoo, and it is an absolute honor to join this team," Large said. "I can't wait to continue the zoo's mission of mixing fun with nature education, helping families create memories and looking forward to our best days ever." She added, "Everyone at home is excited about the zoo and have requested to come to work with me every day."

One of Carrie's main goals was "to bring our neighboring community into the zoo more often and to expand the zoo's connections with people of all walks of life and all abilities." It was she who conceived the idea of adding a Children's Veterinary Clinic to the new Animal Health Center so youngsters could "play vet" with plush animals and toy medical tools based on what they learn on their visits.

Chapter 5

A TWENTY-FIRST-CENTURY ZOO

One Health/One Mission was the most critically important project undertaken at the zoo in over a decade. The Friends' fundraising statement said,

> *The Rosamond Gifford Zoo at Burnet Park is a national leader in animal care, research, conservation and saving species. As stewards of threatened and endangered species, we are dedicated to ensuring the health and well-being of the creatures under our care—matching passion with expertise—and setting the standard for nationally and internationally accredited zoos in the twenty-first century.*

The Friends explained that "the Animal Health Center will also encompass a teaching hospital for future zoo veterinarians, a forum for medical and behavioral research, and public education opportunities for learners of all ages."

The fundraising campaign and case statement were "based on the One Health philosophy that embraces the intrinsic connection between people, animals, plants and our shared environment to achieve optimal global wellness." The campaign included opportunities for individuals and organizations to sponsor a significant space or piece of equipment.

> *For every donor that contributed to the campaign, there seemed to be an inspiring personal story that made these gifts even more meaningful.*

Above: In 2021, construction began for the One Health/One Mission project, which would include a health care center, a teaching hospital, research facilities and public education opportunities.

Left: Ted Fox and Carrie Large show off a display of the Animal Health Center, one of the most important projects ever undertaken at the zoo. The $1.1 million facility would make the RGZ a national leader in animal care.

> *Michael Sperling, who uses oxygen himself, donated $5,000 for an oxygen holding cage at the health center. Barbara Jean Coffey's children honored their late mother's interest in science by raising $10,000 to dedicate the necropsy room in her memory. Longtime zoo supporters Zalie and Bob Linn, whose names grace the Zalie and Bob Linn Amur Leopard Woodland at the zoo, contributed $50,000 to sponsor the Animal Health Center lobby, with its expansive windows looking in on rooms where animal health care happens.*[30]

By the end of 2021, the campaign had raised $890,000 of its $1.1 million goal. Eventually, the campaign surpassed its $1.1 million goal and is still receiving funds to support research and a full-time veterinary position. The Junior Veterinary Clinic, where children explore the role of a vet through play, is a major highlight of guests' visits, with many kids wanting to stay there all day.

Onondaga County covered the $7.5 million cost of design, engineering, construction, infrastructure and landscaping of the Animal Health Center. The Friends offered naming rights to important features of the health center. Friends Director Large said, "This is a tremendous opportunity for Central New Yorkers, especially those involved in health care, to be part of something that will put us on the map when it comes to care of our animals and leading research." The new clinic would have a "separate quarantine facility with four specialized areas to meet the needs of carnivores, primates, aquatic animals and birds."[31] "The new health center will bring the zoo up to standard to maintain its accreditation by the Association of Zoos & Aquariums for many years to come," Fox said. It also served the zoo's partnership with the Cornell University College of Veterinary Medicine.

The decades-long relationship between the zoo and the Cornell University College of Veterinary Medicine had begun in 1997 when a team of experts from Ithaca drove north to operate on the zoo's four-month-old elephant, Mali, to repair an umbilical hernia. General anesthesia and surgical interventions in elephants were very risky, but fortunately, the operation was a success.

> *That high-stakes surgery sparked what has become a 22-year relationship between the Rosamond Gifford Zoo and the Cornell University College of Veterinary Medicine.*
>
> *This formal contract, drawn up by* [Dr. George] *Kollias and the zoo, is mutually beneficial, providing world-class care for endangered species while giving veterinarians and students unprecedented access and training on rare species and conditions. It also enabled Cornell to launch a residency program in zoological medicine, which formed a trifecta alongside Cornell's wildlife health center and exotic pet clinic.*[32]

Future veterinarians and zookeepers might be among the children who "practiced" in the zoo's Junior Veterinary Clinic,

> *a colorful room focused on the zoo's mission to educate. Filled with stuffed animals and play medical equipment, including test tubes, microscopes, and*

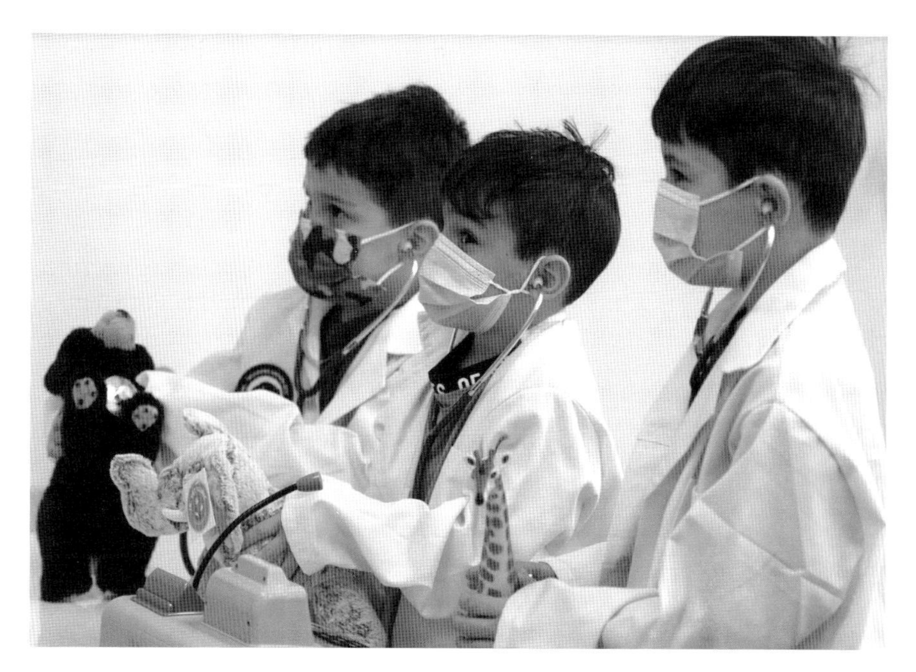

The Junior Veterinary Clinic provides an imaginative play space for learners of all ages and abilities to engage in pretend animal care using toy medical tools and plush animals.

> *scales, this Junior Veterinary Clinic provides an interactive space for little hands and minds to learn about animal care.*
>
> *Stations inside the room replicate a traditional veterinary clinic. Donned in white lab coats provided near the room's entrance, junior veterinarians can visit a waiting room, exam room, surgery station, and kitchen area. At each stop, they can pretend to be a part of the animal care team, performing exams, diagnosing, providing treatment, and more.*[33]

When the zoo closed down during COVID-19, the Friends took advantage of the opportunity to refresh and rebrand the café. In 2022, the HoneyBee Café officially opened—on May 20, World Bee Day. The cafe had a new counter, electronic menu boards and a point-of-sale system, courtesy of the Gifford Foundation. The new menu featured made-to-order sandwiches, salads, pizza, fresh fruit, cookies and more. A coffee bar featured locally roasted coffee beans by Kubal. Online ordering was implemented. In keeping with the zoo's focus on conservation, all food packaging was eco-friendly, with biodegradable cutlery and no plastic straws. The bee theme was intended to highlight a pollinator that is critical to the food chain. Local products were used when possible, such as Beak & Skiff apple cider, Hofman hot dogs and local honey.

Friends director Carrie Large underscored the importance of providing healthier options to visitors. "We want to live our mission in everything we do," she said. "We all live on one planet and need to take care of the environment, the animals, and ourselves. With the One Health, One Mission in mind, we wanted to provide healthier options to enrich our guests, just like we do with the animals." Chuck Anthony, director of food and beverages, added, "Just because the menu is healthier, don't expect it to be boring. The HoneyBee team is invested in providing food that's not only good for you but tastes great, as well."

Catering at the Zoo began to offer "gourmet gatherings for humans that benefit the animals" by creating dining experiences that tell the animals' stories. An arm of the Friends of the Zoo, Catering at the Zoo is "a full-service catering facility focused on organizing and executing spectacular events in a one-of-a-kind setting." It "arranges get-togethers ranging from children's birthday parties to wedding receptions, corporate meetings to company picnics—and everything in between," including private, customized wine/beer pairing experiences for private groups of twenty-five to sixty people.

> *Our beautiful banquet facility can be divided into a smaller banquet room for a more intimate social gathering or opened to create a ballroom space for a large gathering. Our banquet room features brand new wood-grain floors and neutral earth-toned décor that offers a blank slate for any theme or color scheme.*
>
> *Our outdoor patio offers the perfect spot for your cocktail hour. Mix and mingle or break from the party inside. Our patio is surrounded by trees, wooden arbors, flowering plants and hanging Edison lights, making this a unique space and a great spot for one of our ambassador animals to greet your guests.*[34]

The entire zoo can be rented for company picnics.

The zoo launched a highly successful Gourmet Dinner Series that put it on the map of Central New York's best dining experiences and connected diners to the zoo's most important values. April's menu paired the flavors of South Africa with a talk about the creatures that originate from this geographically diverse region, including the zoo's charismatic meerkats and inquisitive Cape porcupines. September's savory seasonal delight, celebrating National Honey Month, paired a mix of ciders, wines and a mead and linked the menu with a discussion by the zoo's beekeepers of the art of beekeeping and the important role of pollinator bees. For Asian Elephant Awareness Month in August, an Elephantastic Eats! grazing menu included seven stations. At Station 1: sweet potato, brussels sprouts, farro, balsamic reduction, goat cheese; Station 2: watermelon, Thai chili, lime, queso fresco, cilantro; Station 3: son-in-law eggs (hard boiled, then fried, with sweet chili sauce); Station 4: sweet and sour tempeh, grilled pineapple, steamed white rice; Station 5: pulled "pork" jackfruit, Korean barbecue sauce, kimchi, brioche roll; Station 6: curry-based chickpea, roasted cauliflower, naan flatbread; Station 7: s'mores. Like all the Zoo's specialty dinners, this one was a sellout.

On the first day of 2022, the zoo made the national news when two male Humboldt penguins became fathers. Bruno's birth on New Year's Day was a good omen for a prosperous year at the zoo.

> *The same-sex foster couple are a first for the zoo, which has relied on foster parents to incubate eggs in the past. The zoo has at least two breeding pairs of penguins with a history of inadvertently breaking their fertilized eggs. To give the eggs a better chance of hatching a chick, keepers may swap a dummy egg for the real one and give it to a more successful pair to incubate.*[35]

"Zoo Director Ted Fox said not all penguin pairs are good at incubating eggs—'It takes practice.'" Such was the case with Poquita and her partner Vente. After Poquita laid an egg and the zoo team detected a viable embryo inside, they decided to transfer the egg to Elmer and Lima, two male penguins who had become a couple in the fall of 2021. Elmer had hatched at the zoo in 2016 and Lima in 2019. "The penguins are free to choose who they want to spend time with (pair with)," Fox explained to CNN, "and in their case, they chose each other."[36] The zoo swapped a dummy egg for the real egg, and Elmer and Lima were given a chance to incubate the real egg. They had built a nest and defended it. They took turns warming the egg, resting it on top of their feet and pushing it up against their feathers. The zoo's marketing director said that "judging from the wave of national and international media attention that Bruno's birth generated, the story of our penguin dads transcended species and cultures, touching hearts around the world. Elmer and Lima's success at fostering is one more story that our zoo can share to help people of all ages and backgrounds relate to animals." The zoo has noted that same-sex penguin pairs show that the idea of "family" is not species-specific and that nontraditional families can do a wonderful job of childrearing.

ONE OF THE DISTINCTIVE elements of the zoo's elephant program is summed up in the motto of the elephant handlers: "Bringing you close enough to care." "Fence time," as the handlers call it, occurs on a number of occasions throughout the day and involves bringing the elephants to zoo visitors. Visitors then have an opportunity to visit with the elephants in a safe environment and ask questions of the handlers. This activity provides visitors to the zoo with a unique opportunity to get close to an elephant and directly experience the awe and wonder those who work with elephants feel on a daily basis. The zoo's elephant handlers are actively involved in the educational and conservation work of the Association of Zoos and Aquariums. The elephant staff's public talks contain an articulate conservation message.

The zoo's elephant herd at the preserve grew to eight in 2022. The new calves were the third generation in the herd's family, which included Mali and her half sister Kirina and Mali's mother, Targa. The herd's composition was like that of natural herds in the wild. The preserve was home to unrelated "aunties," matriarch Siri and Romani. "[The] herd's night house is a 12,000-square foot barn with heated floors and piles of sand to protect their feet." The preserve covers nearly seven acres and includes a fifty-thousand-gallon watering hole that mimics the native habitat of Asian elephants. Asian elephants are slightly smaller than their African cousins and are more critically endangered, with only about thirty thousand remaining in the wild.[37]

The zoo did not start breeding Asian elephants until 1990. Since then, it has had a very successful Asian elephant breeding record and has even held elephants from other facilities for breeding. To date, the zoo has produced seven Asian elephant calves. The elephant management team has four essential aims: to provide the elephants with an environment that is physiologically and socially fulfilling; to increase visitors' and the scientific community's knowledge and understanding of elephants; to support breeding programs in order to contribute to a self-sustaining captive population; and to promote the understanding of conservation issues facing wild elephant populations and to encourage individuals to take action.

Research is an important component of the zoo's mission. On learning that a new COVID-19 vaccine for animals was being distributed to zoos as part of a trial in the summer of 2021, the zoo applied to participate. "The vaccine was developed by Zoetis, a world leader in animal health that has provided vaccines and diagnostic tests for [several other] emerging infectious diseases in animals." After receiving approval from Zoetis, the zoo went through a lengthy state approval system to be part of the trial.[38]

The zoo received enough doses to vaccinate twenty-one animals, including primates, big cats and other mammals, in the first round of the trial. "Thanks to daily training sessions that allow our animals to participate in their own health care," most are accustomed to getting vaccines or allowing blood draws in exchange for their favorite treats and positive reinforcement.[39] All the vaccinations went well, with no ill effects. COVID-19 is a zoonotic illness and is related to other viruses that first appeared in animals; it is believed that the new vaccine will help stop the spread of COVID-19 to other mammals, including domestic cats and dogs, and may even be used in the wild in the form of bait containing the vaccine that wild animals can ingest orally to protect them from the virus.

The zoo's elephant team also spent much of 2021 engaged in important research on behalf of the species in their care, critically endangered Asian elephants. As part of a partnership with the Comparative Cognition for Conservation (CCC) Lab at New York City's Hunter College and the Graduate Center, the elephant herd participated in a new study on how elephants see the world—by smelling it.

Animal behavior researcher Matthew Rudolph of the CCC Lab spent many hours working with the keepers and elephants at the Helga Beck Asian Elephant Preserve to gather data for his study, which asked elephants to detect smaller and smaller amounts of a chemical compound using their noses alone. The team had previously participated in a study of elephant problem-solving using steel puzzle boxes that required the elephants to figure out how to unlock various compartments to find treats.

"When Asian elephant Kirina sees her caregivers wheeling a certain cart toward her stall in the elephant barn, she knows what to do. After months of training for a study of elephant cognition, Kirina knows she'll be asked to play a cognitive game in exchange for her favorite treat, marshmallows."[40] "At the zoo and in the wild, elephants encounter new things every day—new smells, new enrichments, new sounds—and what we learn about how they process this information can help us protect wild elephants," Rudolph explained. "Learning how elephants perceive their world will ultimately help us devise more effective conservation strategies for them in the wild."

Rudolph's work is part of a project led by Dr. Joshua Plotnik and a team of graduate students, who are "studying the evolution of complex cognition across a range of species—from elephants and giant pandas to pangolins and turtles—and asking how behavior and cognition can apply to endangered species conservation."[41]

One of the gravest threats to Asian elephants in the wild is human-elephant conflict in their range countries.

> *For example, farmers in Thailand don't want elephants eating their crops, but the government doesn't want elephants being injured or killed.... Even after parts of their habitats vanish, elephants keep coming back in search of food....Conservationists and farmers in Thailand have tried many methods of "conflict mitigation," including planting barrier crops to block their main crops from view, setting aside easement portions of their crops for elephants, interspersing hot chili peppers, and the old stand-by, electric fencing to keep elephants out. But sooner or later, an elephant will find a way in, and others will follow.*
>
> *"Elephants are great problem-solvers because they're so smart," Matthew said. The CCC Lab began asking: how are elephants solving problems like getting through fences?*
>
> *Our zoo's collaboration with the CCC Lab began with a study on Asian elephant innovation and problem-solving. Ph.D. student Sarah Jacobson designed an elephant-proof puzzle box—a multi-access steel box with compartments the elephants had to figure out how to open to find food.*
>
> *They found that, as Sarah titled her just published paper, "Persistence Is Key"—meaning the most persistent elephants were most likely to innovate to open all the compartments.*
>
> *Some of our elephants behaved predictably for their personalities, said Rosamond Gifford Zoo Elephant Manager Ashley Sheppard. The zoo's eldest Asian elephant, Siri, "would solve the first puzzle right away, but as they got tougher, she would give up," Ashley said, noting that Siri's attitude is, "I like to do things on my terms."*
>
> *Also as expected, food-motivated Kirina would keep trying till she found the treat. But bull elephant Doc surprised the team with his persistence, Ashley said. "We figured that Doc would not be that interested, but he ended up wanting to participate," she said....*
>
> *The upshot is, the most difficult elephants to "mitigate" could be the most persistent.*
>
> *How might this apply to mitigation strategies in the wild? For one thing, it means that relying on the same solutions long-term won't work with the most persistent elephants. "I would say, 'Variety is key,'" Ashley explained. "This research shows that you won't be able to use the same mitigation technique forever. You may have to keep changing it up. Humans have to be persistent, too!"*[42]

The RGZ's research will help conservationists in the range countries of Asian elephants develop better strategies for protecting wild elephants, whose knack for figuring out how to detect agricultural crops and unlock barriers to raid them tends to get them in trouble—and sometimes killed—in places like Thailand. Contributing information on how elephants think will shed light on better ways to mitigate elephant-human conflict in the wild.

RESEARCH AND TRAINING ARE integral parts of the zoo's mission.

> *At the Rosamond Gifford Zoo, our animal care staff work daily to train each animal to participate in their own self-care. Trust is the most essential component of these training programs. Each species requires a unique program with different goals based on the animals' aptitude. For example, primates are more trusting and intelligent than felines who tend to be solitary and suspicious. So, with a cat like Rafferty* [one of the zoo's Amur leopards], *the first goal of the program might be something as simple as getting him to approach the training wall.*
>
> *After rewarding him for accomplishing this, animal staff will direct Rafferty to touch a certain "target" spot on the glass wall of his enclosure. If he does this, he is rewarded with his favorite treat: raw chicken. The training relies on Rafferty's cooperation—if he isn't in the mood for a training session, he doesn't have to participate.*[43]

Pitbull Cora came to the zoo from Helping Hounds Dog Rescue to succeed Moakler as the zoo's elephant dog. Cora has the very important job of teaching the elephant care team how to train an animal before they work with the elephants. The training techniques are the same, but it's a lot easier to work with Cora, who weighs thirty pounds, than with an elephant weighing several tons. The RGZ uses mainly operant conditioning with a heavy emphasis on positive reinforcement. The goal for the animals is to become comfortable with humans. For the humans, the goal is to see how animals learn. The animal care team can practice its techniques with Cora, who is very friendly. Thus she serves as an ambassador for her breed, which is often thought to be vicious, and also for her rescue organization.

The Rosamond Gifford Zoo made history again on October 24, 2022, when twin male Asian elephants were born to Mali and Doc.

> *Mali delivered the first calf at 2 a.m. with no complications. The calf was a male at 220 pounds and perfectly healthy.*
>
> *The second calf came ten hours later at 11:50 a.m. in an event that astonished the animal care staff. The second calf was also male, weighed 237 pounds, but was noticeably weaker than the first. The zoo's animal care team and veterinary staff sprang into action and were able to significantly improve the calf's condition.*[44]

"This is truly an historic moment for the zoo and our community. I couldn't be prouder of our exceptional animal care team, the support of the veterinary staff and their tremendous dedication to Mali and the twins," said county executive Ryan McMahon.

The following dramatic play-by-play description of the twins' birth is based on reporting by Glenn Coin of Syracuse.com.

On October 24, the elephant care team of the Rosamond Gifford Zoo was startled and shocked by the arrival of a newborn elephant calf, who "lay motionless on the floor of the birthing area of the zoo." It was "just seconds after his…arrival into the world," but it was clear that the little

The first successful birth of twin Asian elephants outside their range countries in Asia and Africa at the Rosamond Gifford Zoo in 2022 made headlines.

pachyderm "was in trouble. He made no effort to stand. His heart rate was too slow. His breathing was shallow, his temperature too low. His healthy twin brother had arrived nine hours earlier, but "zookeepers had no idea that [his mother] Mali was carrying twins until the second one slid from her birth canal and landed...on the floor....The elephant manager frantically called zoo Director Ted Fox....'It doesn't look good,' she said. 'He's not responding.'"

> *Mali, 25, had given birth three times before, each to single calves. Keepers wondered during her 22-month pregnancy if Mali might be carrying twins this time. "We joked about this a lot through the last three or four months because she looked so huge," Fox said. An elephant reproduction expert in Florida assured the zoo that elephants who've had multiple pregnancies often look bigger. Keepers don't expect an elephant to bear two calves at once. Twin births are astonishingly rare in the wild and in captivity, happening in less than 1% of pregnancies. When twin pregnancies do occur, they often end in miscarriage, or the twins are stillborn. Only one set of Asian elephant twins has been confirmed in the wild, according to a 2020 study. In captivity, there had been a combined eight twin pregnancies in North American and European zoos. Of those 16 calves, 15 died....*
>
> *Curator Dan Meates says he's often asked why the zoo didn't know Mali had a second, nearly 240-pound calf inside her last October.*
>
> *X-rays can't penetrate an elephant's thick hide and dense muscles, he said. The zoo did routine sonograms during Mali's pregnancy. But the hand-held sonogram probe, about the size of an electric razor, can capture only a small section of the uterus....*
>
> *"You just see a foot or you see a couple of ribs," Meates explained. "You never see a whole image of the baby."*
>
> *Trying to deliver twins can be deadly to the mother. Two calves can get stuck inside the womb and die, said Dennis Schmitt, an elephant reproductive expert in Missouri who has worked with the Syracuse zoo. A farmer can reach inside a cow and reposition a calf; an elephant keeper can't reach inside and maneuver a 200-pound elephant.*
>
> *Schmitt said an elephant's birth canal is "more convoluted" than a horse's or a cow's. An elephant calf emerges not from the back side, like a cow, he said, but from beneath the elephant. An elephant calf must first go up, over the mother's pelvic bone, and then slide down a 6-foot-long birth canal.*

That meant the second calf had lingered inside Mali for nine hours after his brother was born.

The birth of Mali's first calf Oct. 24 had gone smoothly. The 220-pound calf, later named Yaad, dropped about three feet from his mother's birth canal onto the straw-covered, heated floor of the maternity stall at about 2 a.m.

Yaad was soon standing up, shuffling and breathing normally. Mali, who weighs about 9,000 pounds, had taken to her newborn, rubbing and sniffing him with her trunk. He was already suckling.

Mali's mother, Targa, stood in the next stall, able to see and reach her trunk through the bars to the maternity stall. On the other side of the stall were Mali's half-sister, Kirina, and Kirina's mother, Romani. All watched and sniffed intently as Mali and Yaad bonded.

Euphoric but exhausted zookeepers, some who had pulled 48-hour shifts waiting for Mali to go into labor, had one final task: Catch the placenta, which can weigh up to 40 pounds, on clean tarps, whisk it away and send samples to a lab in Michigan. Researchers there grow stem cells from placentas in hopes of curing a devastating elephant virus that had killed two Syracuse calves in 2020 [EEHV, the devastating virus that had killed Ajay and Batu in Syracuse in 2020 and was a killer of many elephants both in zoos and in the wild].

In Mali's three previous pregnancies, the placenta had slid out in two or three hours. This time, the 10 zookeepers waited. And kept waiting. Four hours passed, then five. No placenta. No contractions.... Finally, just before 11 a.m.—nearly nine hours after Yaad was born—Mali started showing mild contractions that suggested the placenta was on its way. Eddy had briefly turned away to ready a tarp.

"The next thing I remember was hearing, 'It's a baby!'" Eddy recalled....

Without warning, a second, an even larger male calf had suddenly whooshed out of Mali's birth canal and dropped to the floor.

"It's very fast," Fox said. "It's like a luge."

Keepers were stunned.

"I do think we're all in shock for maybe like five seconds before we jumped into gear," recalled Seth Groesbeck, the zoo's collection manager. "It was hard to even tell if he was alive when he first came out."

In the adrenaline rush, the bleary-eyed keepers fell back on the training they had conducted during Mali's 652-day pregnancy. Meates checked the calf's temperature; it was too low. Someone warmed bags of saline solution in a microwave and laid them on the calf's side like hot water bottles. They rubbed the calf's body to stimulate circulation....

The staff was ready. During Mali's pregnancy, they had repeatedly practiced how to care for a distressed calf, with a nearly life-sized inflatable adult elephant standing in for Mali. In the drills, a 3-foot long, greased buoy substituted for a newborn calf slippery with amniotic fluid.

As Tukada lay on his side, someone slid a ribbed breathing tube into his mouth and began gently squeezing a bag to get oxygen into his lungs. Eddy suctioned secretions from the calf's trunk. Veterinarians gave him a shot of atropine, which speeds up the heart rate.

After a few minutes, Tukada blinked, raising hope. Tukada showed steady signs of improvement. Within an hour, the frantic efforts of the zoo crew had paid off. Tukada's heart beat more than 80 times a minute. His breathing deepened, his temperature returned to normal.

Keepers knew Tukada had to stand up to get his blood flowing properly. They strained and lifted, and finally he stood, wobbly but upright. Tukada, it finally appeared, had recovered from the traumatic birth.

He wasn't out of the woods yet. Once Tukada's vital signs were stable and he was standing on his own four feet, he faced another hurdle.

"Elephants have evolved to have a single calf," Fox said. "Would the mother accept only the first calf and reject the second, especially if it was compromised and not as strong?"

Holding their collective breath, zookeepers walked Tukada slowly toward Mali.

"She feels him, she smells him," Meates recalled. "And there's vocalization, a lot of vocalization. And a lot of it we can't hear because it's infrasound. They usually flood the babies with infrasound."

Infrasound is a communication mechanism for animals such as elephants, whales, rhinos, giraffes and alligators and has many harmonics, which humans can only hear if they are loud.

Soon it became clear that Mali was accepting her second son. "She's touching the baby, she's smelling it, they're doing trunk-to-trunk exchanges of smells," Groesbeck said. "You can tell by her overall demeanor that she's accepting of the second calf."

But nursing was still an issue. The second twin wasn't understanding the process. "To encourage Tukada to nurse, keepers held the feeding tube next to one [of] Mali's two teats, on her chest between her front legs."

"We got the baby in the position under the mom where it should be, and we used a small rubber tube and inserted that into his mouth while he was trying to nurse on Mom," Groesbeck said.

For a week, keepers fed the calf as many as twenty times a day. At last, he chose to nurse from his mother.

"It took about seven days total for him to fully latch on,' Groesbeck said. "He actually told us he was done with us feeding him. We would try to put the tube in his mouth and he would get upset about it and back off Mom."

Mali produced enough milk for both calves, and they started to eat solid food such as grapes and mashed apples. Both gained weight: almost one hundred pounds within two weeks. The twins were named Yaad (Memory in Hindi) and Takada (Chip).

"I can't commend my team enough for all they have done these past few weeks to ensure the care and safety of Mali and her twins," said zoo director Ted Fox. "It has been incredible to watch them in action and witness the high level of expertise, professionalism and focus under pressure. The continued work and research that follows will significantly contribute to global research efforts on behalf of elephant care."

"I would attribute the success of the birth to the great care that these elephants receive every day," elephant caretaker Alinda Dygert said. "The wild is not a great place for a lot of different animals, Asian elephants in particular, and being able to ultrasound Mali throughout the pregnancy, making sure she had all the food she needed at all times, and also the care that Tukada received after birth were vital to their survival."[45]

Attendance records were broken as thousands came to see Yaad and Tukada. Zoo director Fox said that over forty-two thousand people entered the gates during the two weeks of spring break, and Memorial Day weekend brought in almost three thousand people per day. The twins were available for viewing twice a day at the Helga Beck Asian Elephant Preserve. People came from near and far to see them.

The zoo celebrated the twins' first birthday on October 22, 2023. The Central New York community came together for an afternoon of fun-filled celebrations. "Happy Birthday" was sung to the "EleTwins" before they got to smash a birthday cake filled with vegetables, frozen fruit, whipped cream and cereal. Visitors who wanted to give the twins a birthday present could purchase items for them at the Curious Cub Gift Shop. The twins enjoy snacks like jelly beans and marshmallows, fun toys, engaging puzzles and boomer balls, according to the zoo.[46]

The elephant twins grew as quickly as the attendance figures. At the age of one, each weighed more than 560 pounds, and a permanent home for them became a topic of conversation. "It's normal for adult male elephants to be moved," Fox explained. "We want to follow the natural structure of elephant society as closely as we can, which means that when the bull calves are five or six years old, they would start to create their own social groups." The zoo in Denver was a possible next home, where the twins' older brother Little Chuck lived. "We don't really want to see them go," said Fox. "We have talked about expansion at the elephant preserve area." That project would require funding. Micron Technology, which funded a free day at the zoo for the community, expressed interest in helping.

ANOTHER JOYOUS BIRTH WAS recorded at the zoo in 2023. First-time mother Zeya and father Thimbu became the parents of twins after three years of conservation efforts. Amur tigers, some of the rarest large cats on earth and one of the most endangered species on the planet, are native to the Amur region of China and Siberia. The wild population is estimated to be less than four hundred. Thimbu had come to the zoo in 2019 from Colorado, and Zeya came in 2020 from Connecticut. They had been paired as part of the AZA's Species Survival Plan. Unique to the situation is that Zeya was hand-raised as an infant and it wasn't clear whether she would be a good mother to her cubs. Fortunately, she was an exceptional mother to the litter since their birth. The cubs, one male and one female, were named Zuzaan and Soba.[47]

IN 2023, TWO FEMALE koalas, Kumiri and Kolet, took up temporary residence at the Koala Outpost. The koalas came from the Koala Conservation and Education Loan Program, "which seeks to educate the public and encourage people to care about the species."

> *"Featuring a new species at the zoo is always something to look forward to, but when they are as rare as these koalas, the anticipation becomes exhilarating," said Friends Executive Director Carrie Large. "These koalas give us the chance to fulfill our conservation mission by providing care to an increasingly vulnerable species while educating the public to help connect them to care about all wildlife."*
>
> *"Koalas are rare in American zoos. Only 10 zoos in the United States are permitted to care for koalas by the Australian government and the United States Fish and Wildlife Service," said Executive Director Ted Fox. "The temporary acquisition of these koalas is an important milestone in our zoo's conservation mission and speaks to the expertise and qualifications of our animal care team."*[48]

EDUCATIONAL PROGRAMMING IS ESSENTIAL to the zoo's mission. "Zoo to You is a traveling, inquiry-based educational program that visits schools, libraries, community centers, senior centers/facilities, day care centers, scout groups and more." Zoo to You's goal is to "increase awareness of the natural world and encourage participants to be environmentally conscious." "Each program includes biofacts (animal-related artifacts such as furs, feathers, and skulls) to create a hands-on experience for participants."[49] Zoo Edventures is a three-week learning series for preschoolers. Each themed class includes animal biofacts, nature-themed sensory play, a story and an opportunity to see an animal up close.

THE ENTERTAINMENT ASPECT OF zoos is not missing at the RGZ. It just takes place around the animals without stressing them. The Asian Elephant Extravaganza is an annual summer "celebration of the zoo's eight-member elephant herd and the cultures of their native range countries." One of the zoo's biggest events for the biggest species in its care, Extravaganza attracts thousands of "ele-fans" to an early morning yoga session, followed by games, music and dance performances and elephant activities. Guests enjoy

One of the Rosamond Gifford Zoo's most beloved family events is Zoo Boo, a safe, fun, kid-friendly, "kooky-not-spooky" daytime Halloween celebration.

an elephant watermelon smash, an elephant birthday celebration, keeper chats at the Helga Beck Asian Elephant Preserve and a raffle of animal art pieces. Representatives from Syracuse University's South Asia Center provide "rangoli coloring sheet craft, promoting the Indian art style" and teach visitors how to say "elephant" in the languages used in the regions of India in which elephants live. "Zoo volunteers educate visitors on the importance of choosing sustainable palm oil—explaining how unsustainable palm oil sourcing is harmful to Asian elephant populations."[50]

Zoo Boo is a members-only family event: themed weekends held each October. It offers safe, fun, "kooky-not-spooky" daytime Halloween celebrations as the zoo is transformed into a kid-friendly haunt with trick-or-treat stations, creepy crawly animal encounters, games and activities. Guests are invited to wear costumes and bring treat bags for weekends based on themes such as Superheroes, Witches and Wizards or Favorite Movie or TV Characters. "Mysterious fog and eerie music spill over our decorated trail, giving a unique Halloween feel to visitors without being too scary for the little ones. A trick-or-treating station in the Animal Health Center [has] a photo op where Zoo Boo participants can pose for pictures." There is face painting and a costume parade. Lots of keeper chats round out the fun and excitement while providing important lessons about the animals and about conservation.[51]

The zoo's Holiday Nights provide a "festive and fun after-hours stroll amid sparkling displays of holiday lights that transform the zoo into a winter wonderland." Guests are invited to "warm up by the fire pits and enjoy hot chocolate, s'mores, live performances of holiday music, ice carving, roving entertainers and animals on evening exhibit." In 2023, "in the spirit of the season, the Friends Zoo United campaign pledged to donate $1 from every hot cocoa purchased at Holiday Nights to United Way of Central New York. During Holiday Nights 2023 we served an incredible 176 gallons of hot cocoa and donated $1,876!"[52]

The Halloween Zoo Boo and winter holiday celebrations are tremendously popular, and there are other special events designed for special guests. Dreamnight, an evening dedicated to allowing children with special health care needs and disabilities to experience the zoo, annually allows almost seven hundred people to enjoy a sensory-inclusive after-hours walk, keeper chats, games and various performances. In 2022, "children received a stuffed animal upon their arrival and families were provided dinner and snacks throughout the night." The zoo partners with local organizations to provide free tickets for the fun-filled evening and the organizations provided volunteers to help run it.[53]

The fact that the Rosamond Gifford Zoo is open year-round, embracing each season, also brings some unique challenges. Syracuse, with a population of 100,000 people, is the snowiest U.S. city, averaging just under eleven feet of snow a year from as early as October 1 to as late as May. Operating a zoo with many outdoor exhibits in a climate like this can be challenging in terms of ensuring the safety and comfort of animals and visitors. Director Fox notes that cooler temperatures are "actually enjoyable for a lot of the animals" but that the zoo has "a lot of things in place, from shelters and heated areas and extra feed" in the winter months. "During a blizzard, the animals are given additional blankets to keep them from laying directly on the snow, and they are able to enter heated barns that protect them from the elements.... The pond is equipped with bubblers to maintain motion in the water, but when that isn't enough...zookeepers manually cut and remove ice from the water."[54]

> *In January and February, also known as "Snow Leopard Days," the zoo hosts a variety of events, including gourmet dinners and craft-based activities for children. During these months, admission to the zoo is reduced as a way to incentivize guests to visit. Despite the weather, the zoo aims to draw patrons out and provide fun activities in addition to seeing the animals.*
>
> *"We get questioned a lot if we're open during the winter, because most people don't feel like going outside necessarily, so they don't think our animals would be outside either," said Tammy Singer, a collections manager at the zoo. "Winter is actually a really great time to come to the zoo and see the animals because they're more active, they're more visible and they're not hunkering down in the heat of the day to take a nap."*[55]

"Little Giants" is the nickname of an elephant sculpture monument installed at the zoo in the fall of 2023. The sculpture was created by Ellen Rogers and was donated by Stone Quarry Hill Art Park in Cazenovia to commemorate Asian elephant brothers Ajay and Batu, who succumbed to the deadly elephant endotheliotropic herpes virus (EEHV) in 2020. The zoo is coordinating with national and international partner institutions such as Cornell University and Surrey University to help advance EEHV vaccine research. The sculpture is "a reflection on [Rogers's] experience as a wildlife veterinarian in Africa, where she aided conservation efforts for critically endangered species. Entitled 'Welcome to the Anthropocene,' the abstract artwork is an interpretation of humanity's influence on nature and wildlife." The artist explained, "My artwork addresses the Anthropocene

Epoch, the current geological time period defined by humans causing vast and unprecedented changes to the biosphere: changing climate, massive loss of natural habitat, and animal extinctions." Rogers's elephant sculpture will "help the zoo continue its conservation mission while memorializing Ajay and Batu. Informational signage near the sculpture directs guests to the Ajay and Batu Memorial Fund—allowing visitors to support the zoo's work and join the fight to rebuild Asian elephant populations."[56]

As the zoo prepared to celebrate its 110th anniversary in 2024, it announced that, "awoken from April's solar eclipse, dragons [would descend] on the zoo and begin their reign. Dragons Reign! brings massive, animatronic dragons to the zoo from Memorial Weekend through Halloween. Zoo guests will encounter 10 magnificent 'species' of dragon on exhibit as they explore the zoo. The dragon installations vary in size from five to 20 feet tall."

Carrie Large, executive director of Friends of the Zoo, noted that "dragon mythology is prevalent in many cultures, past and present, around the world, and these dragons reflect both the commonality and diversity of global cultures."

> *Some of the dragons are inspired by different cultures or regions; a Chinese Dragon pays homage to the Zodiac calendar, a pterodactyl-like species called a Quetzalcoatl references the feathered-serpent god of Mesoamerican mythos and folklore, and a Manticore dragon draws on Persian legends. A Wyvern represents the monsters of Anglo-French folklore, while the Western Dragon epitomizes the iconic, classic dragon prominent in Western media.*[57]

Friends of the Zoo planned to host a number of fun-filled themed activities and special events throughout the dragons' reign.

Chapter 6

LOOKING TO THE FUTURE

In 1934, twenty years after the Burnet Park Zoo opened, the following article was published in the local paper:

> *More than 10,000 persons visited the Burnet Park Zoo yesterday to see young Petey Monk, the nine-day-old son of Minnie and Jocko Monk. Attendance records were trebled, according to Daniel Hanley, keeper of the zoo, because of the arrival of the new monkey. While Petey's mother is proud, she is careful that her son does not have too much excitement. Yesterday, with all the crowds at the zoo, Mini kept Petey in a corner.*

We cringe when we read this ninety years later, upset by the anthropomorphism, the caging of the animals and the trauma that the new mother and her infant must have experienced at being stared at by ten thousand people in one day.

The first zoos in the United States were displays of wealth and power. Animals were kept for entertainment and pleasure in private menageries. Public-spirited citizens often wanted to share the exoticism of their collections with the general public. The first incarnation of a zoological park in Syracuse, a small, four-acre facility in Burnet Park with fewer than one hundred animals kept in less-than-ideal conditions, was just that kind of facility. Over a century later, it has evolved and expanded to become a modern twenty-first-century zoo with a state-of-the-art health center and over seven hundred species displayed in the best possible conditions.

And yet, have things changed all that much? Over forty-two thousand people entered the gates of the Rosamond Gifford Zoo during the two weeks following the extraordinary birth of twin elephants. The babies were kept from public viewing for two weeks after being born, were not kept in a cage and had the absolute best in veterinary care. But like Petey the monkey, Yaad and Tukada were objects of intense interest and concern to the public, and three thousand visitors came each day to see the little pachyderms once they were on exhibit.

What did these visitors learn from their zoo experience? What did the zoo teach them? What difference can a zoo make in a world where it has been estimated that 85 percent of wetlands, 75 percent of ice-free land and 63 percent of oceans have been transformed by humans in such a way as to lead to habitat loss, which increasingly threatens the survival of wildlife?

An article in the *Journal of Applied Animal Welfare Science* asked the question, "Where are zoos going—or are they gone?" Author Carl Safina explains that "to some, zoos are prisons exploiting animals. In reality zoos range from bad to better." However, he makes a distinction: "A bad zoo makes animals work for it; a good zoo works for animals." Safina explains,

> *Good zoos do effective conservation work and continually strive to improve exhibits, relevance to conservation, and inspiring public engagement for wildlife. Many zoos have improved enormously; the better ones being crucial in saving species that would have otherwise gone extinct. Nonetheless, for some people the mere word "zoo" carries impressions of old zoos, bad zoos, circuses, and theme-park shows that many find distasteful. Good zoos know they must innovate forward. As society grows increasingly estranged from nature and continues driving broad declines of wildlife, wild lands, and natural systems, the goal of zoos and every organization concerned with animal welfare should not be to separate humans from other animals, but to entangle all humans in nonhuman lives.*

For Safina and like-minded others,

> *Zoos of the next decades must become the first stage in bringing young people into life-long, engaged caring about animals. They could carry on that mission in their communities, in schools, in wild lands, as well as inside their gates. Without a strong public constituency, wild animals will not withstand continued human proliferation. Zoos and aquariums must*

innovate toward being a crucial force abetting the continued existence of wildness on Earth. Zoos of the future must become uplifting places of respect, rescue, enhancement, conservation, and public engagement.

Another article in the same journal asked, "What is the future for zoos and aquariums?" It noted,

Animal welfare concerns have plagued the professional zoo and aquarium field for decades. Societal differences remain concerning the well-being of animals, but it appears a shift is emerging. Scientific studies of animal welfare have dramatically increased, establishing that many previous concerns were not misguided public empathy or anthropomorphism. As a result, both zoo and aquarium animal welfare policy and science are now at the center of attention within the world's professional zoos and aquariums. It is now possible to view a future that embraces the well-being of individual captive exotic animals, as well as that of their species, and one in which professional zoos and aquariums are dedicated equally to advancing both.

The authors, Ron Kagan, Stephanie Allard and Scott Carter, argue that "if animal welfare science and policy are strongly rooted in compassion and embedded in robust accreditation systems, the basic zoo/aquarium paradigm will move toward a more thoughtful approach to the interface between visitors and animals. It starts with a fundamental commitment to the welfare of individual animals."

The Rosamond Gifford Zoo declares with pride:

One of the primary missions of the Rosamond Gifford Zoo and our fellow accredited institutions in the Association of Zoos & Aquariums (AZA) is saving endangered species from extinction. Our 240 top-tier zoos and aquariums collectively approach this mission via a variety of programs to protect and maintain populations of endangered species in our care while also pooling our resources, time and energy to combat the threats that are decreasing their numbers in the wild.

How viable is this mission?

Species Survival Plans (SSPs) are population management programs for specific species. The main goal of SSPs is to maintain genetically diverse,

> *multi-generational, and stable populations of animals in human care. Each plan keeps a carefully recorded studbook and publishes a breeding and transfer plan regularly. Every birth or hatching is carefully planned to improve the species in human care. This helps AZA members manage species collectively, with careful planning for the future of the species both in human care and with hopes of future reintroduction programs.*[58]

The Rosamond Gifford Zoo maintains studbooks containing genetic and breeding history and breeding and transfer plans for five of the five hundred animals in the AZA's SSP programs: the Turkmenian markhor, Armenian mouflon, fairy bluebird, patas monkey and Thorold's white-lipped deer. At the same time, even this exercise is fraught with challenges. The AZA recognizes that many SSPs were not sustainable and would no longer exist in zoos and aquariums if they didn't change their approach.

Elephants have been at the heart of the Syracuse zoo since its inception. Today *Elephas maximus*, the Asian elephant, is endangered. Asian elephant numbers have dropped by at least 50 percent over the last three generations. Where once large herds of elephants roamed freely throughout Asia's forests and grasslands, today the wild population is estimated to number from thirty thousand to fifty thousand, with continuing decreases caused by habitat loss, human/elephant conflict and poaching.

The Rosamond Gifford Zoo is recognized as a leader in the care and husbandry of critically endangered Asian elephants in conjunction with the AZA's SAFE program. SAFE, which stands for Saving Animals From Extinction, was founded in 2015 to leverage the collective expertise of AZA-accredited zoos and aquariums and the size of their base of visitors and guests to save threatened species. Asian elephants are among the SAFE species with dedicated AZA conservation programs aimed at countering habitat loss, fragmentation and degradation of their wild populations.

The principal habitat of Asian elephants is the forest. Human population growth results in the destruction of forests, through "intensive logging, clearing of forested land for agriculture, livestock grazing and infrastructure development for human settlements. As the natural habitat of Asian elephants shrinks, hungry elephants are forced to search for food outside of the forest." They begin to enter into competition with human populations for crops such as bananas, rice and cassava, which results in angry human retaliation. In Sri Lanka, for example, more than 100 elephants and 50 people are killed annually during conflicts. In addition, Asian elephants are poached regularly for their ivory tusks and other body parts.[59]

So what difference can the Rosamond Gifford Zoo make? What has it learned from Siri, Ajay, Batu, Yaad and Tukada and their families that can help Asian elephants in the wild overcome the seemingly inevitable destruction of their species?

Because half of the world's population is urban and lives remote from nature, urban zoos such as Rosamond Gifford have the potential to win support for wildlife preservation through education about animals, their habitats, their importance and their potential destruction and elimination from the planet. This educative function has become increasingly and vitally important to the RGZ.

The zoo is constantly innovating and developing ideas for habitat renovations and expansions. One such idea in the concept stage is an expansion of the snow leopard exhibit on the Wildlife Trail, including the creation of a "snow leopard overpass." Like domestic cats, wild cats love to climb to high elevations to gain a vantage. An overpass would enable the snow leopards to lounge on or walk across an overhead bridge from their existing habitat to a new, expanded exhibit and would offer zoo guests an exciting new perspective on these impressive felines. This exhibit expansion would also incorporate a training area, providing guests with an opportunity to witness animal care staff training snow leopards to participate in their own health care through positive reinforcement.[60]

Robert H. Linn, benefactor and former zoo board chair, wrote an article titled "Dominion Over Creatures" for the local *Jewish Observer* newspaper. In it, he pointed out that in Genesis 1:28, God said to man, "Be fruitful, and multiply, and replenish the earth, and have dominion over the fish of the sea, and over the fowl of the air, and over every living thing that creepeth upon the earth." Linn explained that Jewish scholars have interpreted the word *dominion* to mean that man is not solely to have control over other species "but rather that man should have responsible stewardship over animals and provide for their health, safety and replenishment." Linn went on to note that while there are over two thousand "animal exhibitors" in the United States, there are only 241 accredited zoos. He noted that the AZA accreditation process "is no small task for the AZA and local zoo."

Linn was cognizant of the fact that "many people find zoos to be inappropriately housing animals that should be in the wild" but argued that "zoos can, if appropriately managed, maintained and accredited, play an important role in the education of young and old alike as to the importance of man having, in the biblical tradition, 'dominion' over wildlife and protection of individual species that are endangered in the wild." He quoted I. Richard G. Conway, former director of the Wildlife Conservation Society and president of the AZA, who said, "The justification for removing an animal from the wild for exhibition must be judged by the value of that exhibition in terms of human education and appreciation, and the suitability and effectiveness of the exhibition in terms of each wild creature's contentment and continued welfare."

Linn noted that while some people have the means to view animals in their natural habitat in Asia and Africa, "very few people get to do this." A local accredited zoo significantly enlarges the pool of those who can see, learn about and appreciate the diversity of animal life on our planet. In addition, Linn cited "the importance of zoos in a community to complement other educational and arts organizations that make up the fabric of a community and make a community a good place to live, learn and raise a family." He concluded, "I am proud of our community's efforts to care for endangered animals and to teach respect and stewardship over animals of the earth."

Carl Safina's excellent article in the *Journal of Applied Animal Welfare Science* titled "Where Are Zoos Going—or Are They Gone?" raises the following very salient points:

> *Good zoos know they must innovate forward. As society grows increasingly estranged from nature and continues driving broad declines of wildlife, wild lands, and natural systems, the goal of zoos and every organization concerned with animal welfare should not be to separate humans from other animals, but to entangle all humans in nonhuman lives. Zoos of the next decades must become the first stage in bringing young people into life-long, engaged caring about animals. They could carry on that mission in their communities, in schools, in wild lands, as well as inside their gates. Without a strong public constituency, wild animals will not withstand continued human proliferation. Zoos and aquariums must innovate toward being a crucial force abetting the continued existence of wildness on Earth. Zoos of the future must become uplifting places of respect, rescue, enhancement, conservation, and public engagement.*

In *The Modern Ark*, Vicki Croke wrote that the challenge of the modern zoo is

> *to allow living, breathing animals to inspire wonder and awe of the natural world; to teach us that animal's place in the cosmos and to illuminate the tangled and fragile web of life that sustains it; to open the door to*

> *conservation for the millions of people who want to help save this planet and the incredible creatures it contains. To enrich, enlighten and empower the people who care, so that through huge numbers and sheer willpower we save the beetle and the snail and the alligator along with the panda and the rhino and the condor.*

Her words are appropriate to the history of Syracuse's Rosamond Gifford Zoo. "No one knows exactly what the zoo of the future will look like. But with so much at stake, it is clear we desperately need zoos to help save the diversity of life. The question is not whether the world will have zoos in the future; the question is: will the world have animals?"

But the last words belong to Carrie Large, executive director of the Friends of the Rosamond Gifford Zoo, who makes the case for the zoo's future boldly and emphatically:

> *The fate of the Rosamond Gifford Zoo—and, in a broader sense, the future of animals, our planet, and humanity—rests in all of our hands. Beyond the essential research that our zoo and partner institutions perform every day in search of cures to ravaging diseases (such as elephant endotheliotropic herpes virus, the leading killer of young Asian elephants), zoos play a crucial role in fostering a meaningful connection with animals, prompting us to develop a genuine sense of interest and responsibility. It is imperative that we persist in devising means to motivate our guests, encouraging them to actively contribute to the well-being of our planet through practices such as recycling, gardening, and demonstrating genuine care for the survival of all species. We need to cultivate interest in the natural world and foster true connections between humans and animals. By creating and strengthening these bonds, we will inspire and encourage people to take action—every generation, every demographic.*

NOTES

Introduction

1. Wikipedia, "Rosamond Gifford Zoo."

Chapter 1

2. Fudacz, "Smith Premier."
3. Howard, "Smith Premier 1 Typewriter."
4. Fudacz, "Smith Premier."
5. Smith Corona, "History."
6. Winship, "Flora Bernice Smith."
7. Winship, "Flora Bernice Smith."

Chapter 2

8. Wikipedia, "Rosamond Gifford Zoo."

Chapter 3

9. Croyle, "Throwback Thursday."

Chapter 4

10. Rosamond Gifford Zoo at Burnet Park, "Zoo Team."
11. Rosamond Gifford Zoo at Burnet Park, "Zoo Team."
12. Moriarty, "Elephants Return."
13. Jennings, "Rounding Up."
14. Jennings, "Rounding Up."
15. ZooBorns, "Red Panda Brothers Born."
16. ZooBorns, "Rosamond Gifford Zoo."
17. *Citizen*, "Baby Asian Elephant Born."
18. Onondaga County Parks, "Baby Asian Elephant Born."
19. Rosamond Gifford Zoo, "Rosamond Gifford Zoo Unveils."
20. Rosamond Gifford Zoo, "Rosamond Gifford Zoo Unveils."
21. Friends of the Rosamond Gifford Zoo, *2020 Annual Report.*
22. Rosamond Gifford Zoo, "Animal Health Center."
23. Rosamond Gifford Zoo, "Rosamond Gifford Zoo Opens."
24. Rosamond Gifford Zoo, "Rosamond Gifford Zoo Opens."
25. Rosamond Gifford Zoo, "Animal Health Center."
26. Rosamond Gifford Zoo, "Rosamond Gifford Zoo Opens."
27. Rosamond Gifford Zoo, "Animal Health Center."
28. Rosamond Gifford Zoo, "Rosamond Gifford Zoo Opens."
29. *My Zoo Magazine*, "Carrie Large Joins Friends."

Chapter 5

30. Friends of the Rosamond Gifford Zoo, *2021 Annual Report.*
31. Rosamond Gifford Zoo, "Friends of the Zoo Launches."
32. Roberts, "Zoo University."
33. *My Zoo Magazine*, "Inspiring Veterinarians of Tomorrow."
34. Catering at the Zoo, "Venues."
35. Rosamond Gifford Zoo, "Penguin Chick Hatched."
36. Ravindran, "Same-Sex Penguin Couple."
37. Eisenstadt, "Syracuse Zoo Responds."
38. Rosamond Gifford Zoo, "Rosamond Gifford Zoo Animals."
39. Rosamond Gifford Zoo, "Rosamond Gifford Zoo Animals."
40. Rosamond Gifford Zoo, "Asian Elephant Research."
41. Rosamond Gifford Zoo, "Asian Elephant Research."
42. Rosamond Gifford Zoo, "Asian Elephant Research."

43. My Zoo Magazine, "Specialized Training Sessions."
44. Rosamond Gifford Zoo, "Asian Elephant Delivers Miracle Twins."
45. Weiss, "Significance."
46. Stelk, "'Miracle' Twin Baby Elephants."
47. Rosamond Gifford Zoo, "County Executive McMahon Announces."
48. CNY Central, "Rosamond Gifford Zoo Welcomes."
49. Rosamond Gifford Zoo at Burnet Park, "Zoo to You."
50. Friends of the Rosamond Gifford Zoo, *2023 Annual Report.*
51. Rosamond Gifford Zoo at Burnet Park, "Zoo Boo."
52. Rosamond Gifford Zoo at Burnet Park, "Holiday Nights."
53. Friends of the Rosamond Gifford Zoo, *2022 Annual Report.*
54. Moore, "Even Cold Weather."
55. Moore, "Even Cold Weather."
56. *My Zoo Magazine*, "Modern Art Memorial Sculpture."
57. Rosamond Gifford Zoo at Burnet Park, "Dragons Reign!"

Chapter 6

58. Sculli, "How Are Zoos Helping?"
59. U.S. Fish & Wildlife Service, "Asian Elephant."
60. Friends of the Rosamond Gifford Zoo, *2023 Annual Report.*

BIBLIOGRAPHY

Aiello, J. "Education as a Major Component in Planning a Zoo." *Newsletter of the International Association of Zoo Educators* (1984): 12.

Bonner, Jeffrey P. *Sailing with Noah: Stories from the World of Zoos.* Columbia: University of Missouri Press, 2006.

Catering at the Zoo. "Venues." https://www.cateringatthezoo.org/venues.

Citizen. "Baby Asian Elephant Born at Rosamond Gifford Zoo in Syracuse." Updated January 11, 2020. https://auburnpub.com/news/local/baby-asian-elephant-born-at-rosamond-gifford-zoo-in-syracuse/article_cbb72c12-dd40-5c66-85e7-f6fbc0d8cc50.html.

CNY Central. "Rosamond Gifford Zoo Welcomes Two Koalas." https://cnycentral.com/news/local/rosamond-gifford-zoo-welcomes-two-koalas.

Coin, Glenn. "Inside the Syracuse Zoo's Scramble to Save a Newborn Elephant Twin: 'He's Not Responding!'" Syracuse.com, last updated October 24, 2023.

Croke, Vicki Constantine. *The Modern Ark: The Story of Zoos: Past, Present and Future.* New York: Scribner, 1997.

Croyle, Johnathan. "Throwback Thursday: Syracuse Opens a Zoo to Be 'Proud Of.'" Syracuse.com. Updated September 15, 2023. https://www.syracuse.com/vintage/2016/08/throwback_thursday_syracuse_op.html.

Eisenstadt, Marnie. "Syracuse Zoo Responds to Worst 10 Elephant Zoos List: 'We Are Shocked at the Level of Misinformation.'" Syracuse.com, January 23, 2020. https://www.syracuse.com/news/2020/01/syracuse-zoo-responds-to-worst-10-elephant-zoos-list-we-are-shocked-at-the-level-of-misinformation.html.

Friends of the Rosamond Gifford Zoo. *2020 Annual Report.* https://www.rosamondgiffordzoo.org/assets/Syracuse-Zoo-FOTZ-2020-Annual-Report.pdf.

———. *2021 Annual Report.* https://www.rosamondgiffordzoo.org/assets/Syracuse-Zoo-FOTZ-2021-Annual-Report-05-24.pdf.

———. *2022 Annual Report.* https://www.rosamondgiffordzoo.org/assets/2022-Annual-Report-UPDATED-05-25-23.pdf.

———. *2023 Annual Report.* https://www.rosamondgiffordzoo.org/assets/2023-Annual-Report-CURRENT-v2.pdf.

Fudacz, Greg. "Smith Premier." Antikey Chop. https://www.antikeychop.com/smith-premier-typewriters.

Gray, John M. *Master Plan.* Burnet Park Zoo, 1972.

Hanson, Elizabeth Hanson. *Animal Attractions: Nature on Display in American Zoos.* Princeton, NJ: Princeton University Press, 2004.

Howard, Martin. "Smith Premier 1 Typewriter." Antique Typewriters: The Martin Howard Collection. https://www.antiquetypewriters.com/typewriter/smith-premier-1-typewriter.

Jennings, Zeke. "Rounding Up." Gift Shop Plus, Spring 2016. https://giftshopmag.com/article/roundup-up.

Kagan, Ron, Stephanie Allard and Scott Carter. "What Is the Future for Zoos and Aquariums?" In "Zoos and Aquariums as Welfare Centres: Ethical Dimensions and Global Commitment." Supplement, *Journal of Applied Animal Welfare Science* 21, no. S1 (2018), S59–S70. https://doi.org/10.1080/10888705.2018.1514302.

Klock, Vera E. "The Zoo." Unpublished paper, Syracuse, New York, 1943.

Linn, Robert H. "Dominion Over Creatures." *Syracuse Jewish Observer*, January 2022.

Moore, Donald E., and Charles E. Doyle, "Elephant Training and Ride Operations, Part I: Animal Health, Cost/Benefit and Philosophy." *Elephant* 2, no. 2 (September 8, 1986). https://doi.org/10.22237/elephant/1521731987.

Moore, Sophia. "Even Cold Weather Can't Stop Animals and Guests from Enjoying Rosamond Gifford Zoo." Daily Orange. https://dailyorange.com/2022/02/even-cold-weather-cant-stop-animals-guests-from-enjoying-rosamond-gifford-zoo.

Moriarty, Rick. "Elephants Return to Renovated Quarters at Rosamond Gifford Zoo." March 28, 2021. https://www.syracuse.com/news/2011/03/elephants_return_to_renovated.html.

My Zoo Magazine. "Carrie Large Joins Friends of the Zoo as Executive Director." Winter 2021. https://www.rosamondgiffordzoo.org/assets/Syracuse-Zoo-RGZ-MyZoo-Magazine-Winter-2021.pdf.

———. "Inspiring Veterinarians of Tomorrow." Summer 2022. https://www.rosamondgiffordzoo.org/assets/Syracuse-Zoo-RGZ-MyZoo-Magazine-Summer-2022-06-16.pdf.

———. "Modern Art Memorial Sculpture Honors Our 'Little Giants.'" Fall 2023. https://www.rosamondgiffordzoo.org/assets/FALL-2023-MyZoo-Magazine-FINAL_MEMBERS.pdf.

———. "Specialized Training Sessions Help Animals Participate in Their Own Care." Summer 2023. https://www.rosamondgiffordzoo.org/assets/Syracuse-Zoo-RGZ-MyZoo-Magazine-Summer-2023.pdf.

Onondaga County Parks. "Baby Asian Elephant Born at Rosamond Gifford Zoo in Syracuse, NY." Friday, January 25, 2019. https://onondagacountyparks.com/index.php/news/show/baby-asian-elephant-born-at-rosamond-gifford-zoo-in-syracuse-ny.

Ravindran, Jeevan. "Same-Sex Penguin Couple Become First-Time Dads at New York Zoo." Updated February 2, 2022. https://www.cnn.com/travel/article/gay-penguins-hatch-egg-scli-intl-scn/index.html.

Roberts, Lauren Cahoon. "Zoo University: Education and Conservation Drive Cornell-Zoo Partnership." *'Scopes* no. 1 (2019). https://www.vet.cornell.edu/scopes/issues/2019-issue-1/zoo-university.

Rosamond Gifford Zoo at Burnet Park. "Animal Health Center." https://www.rosamondgiffordzoo.org/experience/animal-health-center.

———. "Asian Elephant Delivers Miracle Twins at Rosamond Gifford Zoo." https://rosamondgiffordzoo.org/news/zoo-news/asian-elephant-delivers-miracle-twins-at-rosamond-gifford-zoo.

———. "Asian Elephant Research: A Study in Innovative Problem Solving." https://www.rosamondgiffordzoo.org/conservation/research-at-the-zoo/asian-elephant-research-a-study-in-innovative-problem-solving.

———. "County Executive McMahon Announces Amur Tiger Zeya Gives Birth to Cubs." https://rosamondgiffordzoo.org/news/zoo-news/county-executive-mcmahon-announces-amur-tiger-zeya-gives-birth-to-cubs.

———. "Dragons Reign!" https://www.rosamondgiffordzoo.org/zoo-events/upcoming-events/dragons-reign.

———. "Friends of the Zoo Launches Animal Health Center Capital Campaign." March 15, 2021. https://www.rosamondgiffordzoo.org/news/zoo-news/friends-of-the-zoo-launches-animal-health-center-capital-campaign.

———. "Holiday Nights." https://www.rosamondgiffordzoo.org/zoo-events/upcoming-events/holiday-nights.

———. "Penguin Chick Hatched by Same-Sex Pair at Rosamond Gifford Zoo." https://rosamondgiffordzoo.org/news/zoo-news/penguin-chick-hatched-by-same-sex-pair-at-rosamond-gifford-zoo.

———. "Rosamond Gifford Zoo Animals Participate in COVID Vaccine Trial." https://rosamondgiffordzoo.org/conservation/research-at-the-zoo/rosamond-gifford-zoo-animals-participate-in-covid-vaccine-trial.

———. "Rosamond Gifford Zoo Opens New Animal Health Center." https://www.rosamondgiffordzoo.org/news/zoo-news/rosamond-gifford-zoo-opens-new-animal-health-center.

———. "Rosamond Gifford Zoo Unveils New Habitat for Amur Leopards." September 24, 2020. https://www.syracuse.com/living/2020/09/rosamond-gifford-zoo-unveils-new-habitat-for-amur-leopards.html.

———. "Zoo Boo." https://www.rosamondgiffordzoo.org/events/view/zoo-boo27257.

———. "Zoo Team." https://www.rosamondgiffordzoo.org/about-the-zoo/zoo-team.

———. "Zoo to You." https://www.rosamondgiffordzoo.org/learn/for-educators-and-groups/zoo-to-you.

Rothfels, Nigel. *Savages and Beasts: The Birth of the Modern Zoo.* Baltimore, MD: Johns Hopkins University Press, 2002.

Safina, Carl. "Where Are Zoos Going—or Are They Gone?" In "Zoos and Aquariums as Welfare Centres: Ethical Dimensions and Global Commitment." Supplement, *Journal of Applied Animal Welfare Science* 21, no. S1 (2018), S4–S11. https://doi.org/10.1080/10888705.2018.1515015.

Sculli, Peter. "How Are Zoos Helping to Sustain Wildlife?" Potter Park Zoo, February 11, 2022. https://potterparkzoo.org/sustain-wildlife.

Smith Corona. "History." https://www.smithcorona.com/history.html.

Stelk, Madeline. "'Miracle' Twin Baby Elephants at Rosamond Gifford Zoo Celebrate Their First Birthday." NCC News, 2023. https://nccnews.newhouse.syr.edu/miracle-twin-baby-elephants-at-rosamond-gifford-zoo-celebrate-their-first-birthday.

Stroup, Adrienne. "A Comparison of the Utica Zoo, the Rosamond Gifford Zoo and the National Zoo." Maymester Paper, June 5, 2012.

U.S. Fish & Wildlife Service. "Asian Elephant." https://www.fws.gov/species/asian-elephant-elephas-maximus.

Weiss, Noah. "The Significance of Rosamond Gifford Zoo's Asian Twin Elephants." NCC News, 2023. https://nccnews.newhouse.syr.edu/the-significance-of-rosamond-gifford-zoos-asian-twin-elephants.

Wikipedia. "Rosamond Gifford Zoo." https://en.wikipedia.org/wiki/Rosamond_Gifford_Zoo.

Winship, Kihm. "Flora Bernice Smith of East Lake Road." Skaneateles (blog), January 11, 2013. https://kihm6.wordpress.com/2013/01/11/flora-bernice-smith.

ZooBorns. "Red Panda Brothers Born at Rosamond Gifford Zoo." September 25, 2018. https://www.zooborns.com/zooborns/2018/09/red-panda-brothers-born-at-rosamond-gifford-zoo.html.

———. "Rosamond Gifford Zoo." https://www.zooborns.com/zooborns/rosamund-gifford-zoo.

ABOUT THE AUTHOR

Barbara Sheklin Davis is an educator and author with eclectic interests. She has a bachelor's degree from Barnard College and a master's degree and PhD from Columbia University, where she specialized in Spanish literature of the Golden Age. After teaching Spanish for twenty-five years at Onondaga Community College (OCC), she began a career in Jewish education, heading the Syracuse Hebrew Day School (SHDS) for twenty-seven years. She published several books while at OCC (*Syracuse African Americans* in 2010 and *The Syracuse Jewish Community* in 2012, both with Arcadia Publishing) and several while at SHDS (*100 Jewish Things to Do Before You Die*, Pelican Publishing, 2016; *Two Jews, Three Opinions*, Wipf & Stock, 2018; and *A Parallel Universe*, Hadassah Press, 2018). She retired to take care of her husband who had Parkinson's disease and wrote two more books (*Advice from a Parkinson's Wife: 20 Lessons Learned the Hard Way*, Parker-Hayden, 2019; and, sadly, *Advice from a Parkinson's Widow: 20 Lessons I Never Wanted to Learn*, Parker-Hayden, 2021). Since 2019, she has been working for the Jewish Federation of Central New York and serving as editor of the monthly *Jewish Observer*. Turning to organizations that benefit their communities, she authored two more books with Arcadia: *Symphoria: The Orchestra of Central New York*, in 2021, and *Rosamond Gifford Zoo at Burnet Park*. Barbara is inspired by people who are doers and achievers and who try to make the world a better place, loves to tell their stories, loves to learn and loves to write. She lives in Syracuse, New York, and has three wonderful children and nine absolutely perfect grandchildren.